D0781676

ROYAL SOCIETY OF CHEMISTRY

## Problem solving in analytical chemistry

Compiled and developed by Karen Crawford and Alan Heaton

Edited by Denise Rafferty and Sara Sleigh

Designed by Imogen Bertin and Sara Roberts

Cover photograph taken at Naas Racecourse, Republic of Ireland, by Peter Mooney

Published by the Education Division, The Royal Society of Chemistry

Printed by The Royal Society of Chemistry

For further information on other educational activities undertaken by the Royal Society of Chemistry write to:

The Education Department
The Royal Society of Chemistry
Burlington House
Piccadilly
London W1V 0BN
Email: education@rsc.org

ISBN 1 870343 46 8

British Library Cataloguing in Publication Data.

A catalogue for this book is available from the British Library.

# Problem solving in analytical chemistry

Karen Crawford and Alan Heaton
Marjorie Cutter Scholarship

The Royal Society of Chemistry
1996–1997

# Contents

# Foreword

Problem solving is a fundamental and distinguishing human skill. Every day we solve a myriad of problems in the pursuit of life and happiness. Problems in commerce and industry cover a wider range and usually have greater complexity, and to be competitive, acceptable and economically viable solutions must be found speedily. It is the use of knowledge, not its possession, that conveys advantage. The modern employee must be able to first define problems, and then seek solutions by acquiring and manipulating information. Whilst logic is the workhorse of problem solving, success can also come from creative leaps in imagination.

I commend this book to students of chemistry. It seeks to develop a very important skill. A skill that not only has considerable value in employment, but will also enhance successful living.

Professor J Beacham

# Background

The Royal Society of Chemistry's guidelines – *Design and delivery of degree courses in chemistry leading to professional membership* – emphasise the importance of problem solving in chemistry degree courses. Alan Heaton proposed that the then Higher Education Sub-Committee could assist university and college lecturers in teaching this skill by producing a bank of tried and tested problems.

In order to build on existing good practice, university chemistry departments were invited to submit suitable problems – those collected ranged from simple examination questions to open ended industrial problems, across all areas of chemistry.

The members of the sub-committee met several times to develop the project strategy and assess the suitability of individual problems. They soon realised, however, that more than voluntary, part time efforts were needed if the project was to move forward satisfactorily over a reasonable timescale. A successful application for funding was made to the Society's Marjorie Cutter Bequest, enabling Karen Crawford to be employed full time during 1996 at Liverpool John Moores University. Professor Alex Johnstone from University of Glasgow also collaborated on the project, and offered valuable advice on the theory behind, and approach to, problem solving.

# Introduction

Fifty-five problems in the broad area of analytical chemistry, suitable for first and second year students on three year full time degree courses in chemistry, are presented in this book. Each problem is accompanied by information on suitability – *eg* tutorial, group work and size of group; timescale; and level (first or second year). Skills and knowledge required and acquired are stated; and a full teacher's guide and solution accompany each problem. Problems have been trialled and refined before publication. It is worth emphasising that almost all of the problems are suitable for group work providing students with valuable experience of problem solving in teams.

The comments which follow this introduction, provided by Professor Alex Johnstone, investigate how to approach learning through problem solving. We hope that it will encourage you to think about problem solving and how to teach this skill within a chemistry degree course. The problems have been classified according to the eight types shown on p. vi. This list appears on p. x. We hope that you find this collection of problems very helpful and would welcome your comments.

We are greatly indebted to four groups of people who have assisted us with this project. Firstly, the Society's Higher Education Sub-Committee which made the initial progress on the project. Secondly, those people who submitted the problems included in the book. Thirdly, those people who conducted the trials of the problems and sent us valuable feedback, and lastly Dr Neville Reed and particularly Dr Denise Rafferty at the Royal Society of Chemistry for their contribution to the successful conclusion of the project.

This book represents the conclusion of what we see as the first stage of the project. However, before advancing we require feedback and comment from you, the user of the problems, on our first collection. Please send this, whether it be constructive, critical or complimentary, either to Dr Denise Rafferty at the Society's Education Department or Dr Alan Heaton at Liverpool John Moores University.

Alan Heaton
Karen Crawford

# Learning through problem solving

*"The defining value for students is to learn about learning, to transform themselves from individuals who need to be taught into intellectuals who are first learners, and then self-learners. Teachers are more expert learners whose understanding about how to learn… is what students need at least as much as they do the factual information."*

*(B P Coppola and D S Daniels)*

This book of problems is not just a compilation of problems set in various universities for chemistry undergraduates, but is designed to provide problems, classified under problem types, to give university teachers a resource for teaching the art and craft of problem solving in general.

The students' view of problems is naturally, "How do I solve THIS ONE?". The teachers' approach is (or should be) to help students to see problem solving as a set of generally applicable techniques which can be learned.

However, research evidence in the field of problem solving indicates that merely solving one problem after another does not necessarily breed habits of good problem solving. There are techniques to be learnt by which problem solving becomes more efficient more quickly.

We do not expect chemical knowledge to be taken in by students by some osmotic process and we should not expect problem solving techniques to be absorbed by diffusion.

Each problem in this book is a chemical vehicle for conveying good factual chemistry and provides a means of systematically teaching problem solving skills.

## Types of Problem

To place the problems in a clear framework the variety of problem types in this book are set out in Table 1. These can be compared with the "normal" problems usually encountered (types 1 and 2). Problems of type 3 would involve the student saying "If you want me to do this, I shall need the following…".

| Type | Data | Methods | Outcomes/Goals | Skills bonus |
|---|---|---|---|---|
| 1 | Given | Familiar | Given | Recall of algorithms. |
| 2 | Given | Unfamiliar | Given | Looking for parallels to known methods. |
| 3 | Incomplete | Familiar | Given | Analysis of problem to decide what further data are required. |
| 4 | Incomplete | Unfamiliar | Given | Weighing up possible methods and then deciding on data required. |
| 5 | Given | Familiar | Open | Decision making about appropriate goals. Exploration of knowledge networks. |
| 6 | Given | Unfamiliar | Open | Decisions about goals and choices of appropriate methods. Exploration of knowledge and technique networks. |
| 7 | Incomplete | Familiar | Open | Once goals have been specified by the student the data are seen to be incomplete. |
| 8 | Incomplete | Unfamiliar | Open | Suggestion of goals and methods to get there; consequent need for additional data. All of the above skills. |

**Table 1 Classification of problems**

Type 4 could be exemplified by "How many copper atoms are there in a 2p coin?". This would involve a reasoning chain like: "If I knew the mass of the coin and if I assumed that it was pure copper and if I had the atomic mass of copper and Avogadro's number, I could get an answer, but it would only be approximate. If I do not have a balance and only have a ruler, I could get its volume (approximately) and if I knew the density of copper, I could get a good estimate." This is very different reasoning from that used in types 1 and 2.

Type 5 is much more open and the student is left to judge what would constitute a reasonable answer, for example, "Given the formula $[Co(NH_3)_4Cl_2]$ deduce from it as much as you can". This could yield a range of responses including the oxidation state of the cobalt ion and its d configuration, the name of the complex, its percentage composition, its isomers, its likely reactions and so on.

Type 7 would require the students to specify goals; to achieve these, extra data would have to be requested.

Type 8 might involve giving the students a substance and asking them to suggest uses for it. The students would have to ask for, or find out experimentally, its properties before deciding on uses.

Type 6 would be similar to type 8 but the given substance would be familiar to the students.

In types 3–8 we are trying to exercise thinking skills, encourage flexibility, branching and creativity. We do not denigrate types 1 and 2 but feel that we are short-changing our students if we do not also expose them to other problems.

## General Strategies

In the research literature on problem solving five strands or approaches constantly appear.

### 1. Identify the problem
This is probably the most difficult part, but the table above should help students to identify where the real problem lies. Is there a clear goal? Do I have all the data I need? Have I seen one like this before? This stage is not to be rushed and time spent on it is amply repaid. Sometimes brainstorming is the way into the problem.

### 2. Representing the problem
Most people are visual thinkers and solve problems best when the representation is pictorial or structural. Students should be encouraged to draw structures, no matter how tentative, and then modify them in the light of later data. There are several examples in the book which illustrate this technique very well.

### 3. Selecting the strategy
Sometimes the strategy (or method) is well known. For example, recall a formula, plug in the data and turn the handle (Type 1 in the table above). This is the algorithmic method which computers and students can be programmed to solve.

However, if the method is unfamiliar 'trial and error' can take over as a highly inefficient means. Students then have to be taught to break the problem into simpler parts, some of which will be familiar. When the sub-solutions are applied to an apparently intractable problem, its complexity is reduced and it becomes possible to find a solution.

### 4. Implementing the strategy
This can often be achieved by allocating sub-problems and allowing students to contribute to the general discussion periodically by supplying solutions to the simpler problems they have solved.

### 5. Evaluating solutions
It is good to check the answer for reasonableness but this evaluation can go much further to helping students to become general problem solvers.

The important activity here is to ask the question "How did we go about solving it?". A review of a successful route illuminates and reinforces general problem solving skills.

"What type of problem was it? algorithmic?, brainstorming?, trial structure modified?" "Was team working used to do the problem as a whole or did we do sub-problems?" Questions of this kind are powerful mind-clearing techniques, which have been shown to facilitate the learning of general problem solving skills.

## The Expert and the Novice

The quotation at the beginning of this chapter rightly points out the need for the expert to share with the novice not only his knowledge, but also his ways of acquiring and handling knowledge.

It is estimated that it takes about 5 years or 10 000 hours worth of working in an area to gain full expertise and there seems to be no substitute for the sheer hard work this entails in gaining knowledge and skills. This bank of expertise can then be drawn upon to make us, as teachers, experts in our field and in problem solving within it.

There is good evidence that problem solving is, to a large extent, tied into specific knowledge bases, but there are generalisable aspects. Most chemistry teachers will have observed how poorly students, in a chemistry context, transport their mathematical skills, which they may exhibit adequately in the mathematics class. Students have to be helped to make these skills transportable from one discipline to another and even transportable within one discipline.

The teacher cannot expect the wealth of five years of knowledge to be exhibited by an undergraduate, even in the final year!

Most professional chemists, in retrospect, will admit that they only "put their chemistry together" years after graduation and especially when called upon to teach.

If we can go some way to helping students to accelerate from their novice state towards our expert state, we shall have done them a service.

This book is commended to teachers as a compilation of problems of rich variety, each providing a vehicle for teaching, overtly and systematically, the general skills of problem solving.

Professor A H Johnstone
Centre for Science Education
University of Glasgow

# Classification of problems

**Type 1**
1, 3, 4, 6, 7, 8, 11, 22

**Type 2**
2, 5, 9, 14, 15, 25, 29, 30, 34, 36, 37, 38

**Type 3**
10, 17, 18, 19, 21, 23, 28,  32, 33, 39

**Type 4**
12, 13, 16, 24, 26, 31, 35, 42

**Type 5**
45

**Type 6**
20, 40, 41, 46, 48, 52

**Type 7**
49, 51, 53, 55

**Type 8**
27, 43, 44, 47, 50, 54

# Acknowledgements

We are most grateful to the Higher Education Sub-Committee who collected the initial bank of problems from university chemistry departments:

Dr Will Bland, Kingston University
Dr John Carnduff, Glasgow University
Professor Peter Elliott, ex Nottingham Trent University
Professor Keith Smith, University of Wales, Swansea
Dr Alan Thompson, University of Manchester
Dr Noel Weston, Reading College of Technology

We are grateful to the following colleagues for assisting us with the teacher's guide, and for proofreading the final set of problems:

Dr Ian Bradshaw;
Dr David Johnstone;
Dr Harry Morris;
Dr Barry Nicholls;
Dr Jack Pearce;
Dr Alex Wood; all from Liverpool John Moores University; and
Dr Elizabeth Prichard, Laboratory of the Government Chemist

We are also grateful to the following for sending us problems, or ideas for problems:

Dr Eddie Allen, University of Portsmouth
Dr Jeremy Andrew, Unilever Research, Port Sunlight, Wirral
Dr Jim Ballantine, University of Wales, Swansea
Dr Dave Bannister, Manchester Metropolitan University
Dr Bill Bentley, University of Wales, Swansea
Mrs Alison Bretnall, Bristol Myers Squibb Pharmaceutical, Wirral
Dr Tom Brown, University of Southampton
Dr Stephen Breuer, Lancaster University
Dr Thomas Cowen, Bristol Myers Squibb Pharmaceutical, Wirral
Dr Andy Gray, Solvay Interox Ltd, Widnes
Dr Allen Millichope, Unilever Research, Port Sunlight, Wirral
Dr Cedric Mumford, University of Wales Institute, Cardiff
Dr Colin Peacock, Lancaster University
Dr Andy Platt, Staffordshire University
Dr John Smith, University of Newcastle upon Tyne
Professor Graham Williams, University of Wales, Swansea
Dr Jim Wood, University of Huddersfield
Dr Chris Wormald, University of Bristol
Department of Chemistry, Glasgow University
Wentworth College, University of York

Thanks are due to a number of colleagues for trialling the problems throughout the project, particularly:

Dr Phil Brown, Heriot-Watt University
Dr Clive Buckley, North East Wales Institute of Higher Education
Dr Simon Duckett, University of York

Dr Paul Heelis, North East Wales Institute of Higher Education
Professor John Mann, University of Reading
Dr Kevin Markland, De Montfort University
Dr David Miller, Aston University
Dr Karen Moss, Nottingham Trent University
Dr Brian Murphy, Dublin City University
Dr Andy Platt, Staffordshire University
Dr Brian Plunkett, University of Portsmouth
Dr John Smith, University of Newcastle upon Tyne
Dr Paul Walton, University of York
Dr John Wheeler, Staffordshire University

Thanks also to the Teaching and Learning Service at Glasgow University.

Karen Crawford
Alan Heaton

# Abbreviations

| | |
|---|---|
| AAS | Atomic Absorption Spectroscopy |
| Abs | Absorbance |
| $^{13}$C NMR | Carbon 13 Nuclear Magnetic Resonance |
| CRM | Certified Reference Material |
| DBE | Double Bond Equivalent |
| EDTA | Ethylenediaminetetraacetic acid |
| FTIR | Fourier Transform Infrared Spectroscopy |
| GC–ECD | Gas Chromatography – Electron Capture Detector |
| GC–FID | Gas Chromatography – Flame Ionisation Detector |
| GC–MS | Gas Chromatography – Mass Spectrometry |
| GLC | Gas Liquid Chromatography |
| GLP | Good Laboratory Practice |
| HC | Hydrocarbon |
| $^{1}$H NMR | Proton Nuclear Magnetic Resonance |
| HPLC | High Performance Liquid Chromatography |
| ICP/MS | Inductively Coupled Plasma/Mass Spectrometry |
| ICP/OES | Inductively Coupled Plasma/ Optical Electron Spectroscopy |
| IR | Infrared Spectroscopy |
| K | Kelvin |
| LIMS | Laboratory Information Management System |
| MS | Mass Spectrometry (or Spectrum) |
| NMR | Nuclear Magnetic Resonance |
| PAH | Polycyclicaromatic Hydrocarbon |
| PCB | Polychlorinated biphenyl |
| PID | Photoionisation Detector |
| $R_f$ | Retardation Factor |
| SRM | Standard Reference Material |
| UV | Ultraviolet spectroscopy |

# List of problems

**Each exercise contains the problem for the student (denoted by S) and a tutor's guide (denoted by T)**

# S1

# Calculating concentrations

**Problem**

Calculate the final concentrations in mol dm$^{-3}$ of H$^+$, Na$^+$, Cl$^-$ and SO$_4^{2-}$ when the following three solutions are mixed together:

- 1000 cm$^3$ of 0.10 mol dm$^{-3}$ HCl
- 500 cm$^3$ of 0.20 mol dm$^{-3}$ NaCl
- 500 cm$^3$ of 0.20 mol dm$^{-3}$ Na$_2$SO$_4$

**Prior knowledge**
Basic calculations.

**This problem is suitable for:**
- First year students
- A tutorial (groups of 2–4 students) or class discussions
- Approximately 15 minutes

**Knowledge/skills gained**
Improved calculation skills.

# Calculating concentrations

This problem could be used as a preliminary exercise to test a student's ability.

The problem could be solved by first calculating the total volume. This may cause some difficulties with the students because the information for this is at the end of the problem and a lot of students start solving a problem before reading the entire question through. This is an important technique for successful problem solving, since a number of problems are best worked out by working backwards. The students should recall that all the molecules will fully dissociate into ions as the species are strong electrolytes.

**Solution**

Total volume = 2 dm$^3$ (*ie* 2000 cm$^3$)

All species are strong electrolytes and fully dissociate into aqueous ions.

Final solution contains:  0.05 mol dm$^{-3}$ HCl
0.05 mol dm$^{-3}$ NaCl
0.05 mol dm$^{-3}$ Na$_2$SO$_4$

*ie*  0.05 mol dm$^{-3}$ H$^+$ and 0.05 mol dm$^{-3}$ Cl$^-$
0.05 mol dm$^{-3}$ Na$^+$ and 0.05 mol dm$^{-3}$ Cl$^-$
0.10 mol dm$^{-3}$ Na$^+$ and 0.05 mol dm$^{-3}$ SO$_4^{2-}$

Thus  [H$^+$]                                 = 0.05 mol dm$^{-3}$
[Cl$^-$]     = 0.05 + 0.05  = 0.10 mol dm$^{-3}$
[Na$^+$]    = 0.05 + 0.10  = 0.15 mol dm$^{-3}$
[SO$_4^{2-}$]                              = 0.05 mol dm$^{-3}$

## S2

# Calculations with sulfuric acid

**Problem**

1. How much of a 0.10 mol dm$^{-3}$ H$_2$SO$_4$ solution is required to neutralise 1 g of NaOH? Express your answer in cm$^3$.

2. How would you prepare 100 cm$^3$ of a 0.02 mol dm$^{-3}$ H$_2$SO$_4$ solution from a 0.10 mol dm$^{-3}$ H$_2$SO$_4$ solution?

**Prior knowledge**
Basic calculations.

**This problem is suitable for:**
■ First year students
■ A tutorial – group work (2–3 students)
■ Approximately 20 minutes

**Knowledge/skills gained**
Revision of basic mole calculations.

T2

# Calculations with sulfuric acid

This problem is useful for pre-laboratory work.

**Solution**

1. $2NaOH + H_2SO_4 \rightarrow Na_2SO_4 + 2H_2O$

   molar mass of NaOH is 40 g mol$^{-1}$

   1 mole of $H_2SO_4$ reacts with 2 moles of NaOH
   *ie* 10 dm$^3$ of 0.10 mol dm$^{-3}$ $H_2SO_4$ solution react with 80 g NaOH.
   Therefore, 1 g NaOH reacts with (10 dm$^3$/80) $H_2SO_4$ = 0.125 dm$^3$
   = 125 cm$^3$ of 0.10 mol dm$^{-3}$ $H_2SO_4$

2. Acid must be diluted by a factor of (0.10/0.02) = 5
   Therefore, to prepare a 100 cm$^3$ solution of 0.02 mol dm$^{-3}$ $H_2SO_4$ add
   '1 volume' of acid to '4 volumes' of water
   *ie* pipette 20 cm$^3$ of the acid into a 100 cm$^3$ volumetric flask and make up to
   the mark with water.

# Calculating equations – the reaction between bromate and hydrazine

**Problem**

Exactly 1.00 g of hydrazine sulfate ($N_2H_5^+HSO_4^-$) was dissolved in water and made up to 250 cm$^3$. A 25.00 cm$^3$ sample of this solution was acidified with dilute sulfuric acid and 50.00 cm$^3$ of potassium bromate(V) solution was added. A quantitative reaction took place with the evolution of a colourless gas.

The excess potassium bromate(V) was determined as follows:
An excess of potassium iodide was added and the iodine released was equivalent to 19.50 cm$^3$ of standard sodium thiosulfate solution (0.10 mol dm$^{-3}$).

A 25.00 cm$^3$ sample of the same potassium bromate solution was acidified and treated with excess potassium iodide. The iodine released was equivalent to 25.15 cm$^3$ of the same standard sodium thiosulfate solution.

The following reactions occur quantitatively:

$$BrO_3^- + 6H^+ + 6I^- \rightarrow Br^- + 3H_2O + 3I_2$$
$$2S_2O_3^{2-} + I_2 \rightarrow S_4O_6^{2-} + 2I^-$$

1.  Calculate the concentration of the potassium bromate solution.

2.  Calculate the amount of hydrazine that reacts with one mole of bromate.

3.  Write a balanced equation for the reaction between $N_2H_5^+$ and $BrO_3^-$ under acidic conditions and hence identify the colourless gas evolved.

**Prior knowledge**
Assessing information, devising equations, mole calculations and balancing equations.

**This problem is suitable for:**
■  Second year students
■  A tutorial or class discussion – group work (2–3 students)
■  Approximately 1 hour

**Knowledge/skills gained**
Experience doing calculations – sifting through large amounts of data.
Communication and teamwork skills.

# Calculating equations – the reaction between bromate and hydrazine

**Solution**

1.  Calculating the concentration of the potassium bromate (V) solution. It can be deduced from the equations given that:

    $$BrO_3^- \equiv 3I_2 \equiv 6S_2O_3^{2-}$$

    Therefore,
    concentration of $BrO_3^-$ = 25.15 cm$^3$ x 0.10 mol dm$^{-3}$/25 cm$^3$ x 6
    = 1.68 x 10$^{-2}$ mol dm$^{-3}$

2.  Calculating the amount of hydrazine that reacts with one mole of bromate. Molar mass of hydrazine sulfate is 130.12 g mol$^{-1}$

    25.00 cm$^3$ of $N_2H_5^+$ solution contains 0.10 g/130.12 g mol$^{-1}$
    = 7.69 x 10$^{-4}$ mol of $N_2H_5^+HSO_4^-$

    50.00 cm$^3$ of 1.68 x 10$^{-2}$ mol dm$^{-3}$ $BrO_3^-$ solution was added, therefore
    Amount        = 1.68 x 10$^{-2}$ mol dm$^{-3}$ x (50/1000) dm$^3$
                  = 8.40 x 10$^{-4}$ mol

    19.50 cm$^3$ of 0.10 mol dm$^{-3}$ $S_2O_3^{2-}$ solution were needed, therefore
    Amount        = 0.10 mol dm$^{-3}$ x (19.50/1000) dm$^3$
                  = 1.95 x 10$^{-3}$ mol

    1 mole $BrO_3^- \equiv$ 6 moles $S_2O_3^{2-}$. Therefore, the amount of excess $BrO_3^-$ can be calculated:

    1.95 x 10$^{-3}$ mol/6    = 3.25 x 10$^{-4}$ mol

    Hence $BrO_3^-$ consumed by the reaction with $N_2H_5^+$

                  = 8.40 x 10$^{-4}$ mol - 3.25 x 10$^{-4}$ mol
                  = 5.15 x 10$^{-4}$ mol

    1 mole $BrO_3^-$        $\equiv$ 7.69 x 10$^{-4}$ mol/5.15 x 10$^{-4}$ mol $N_2H_5^+$
                  $\equiv$ 1.49 mol $N_2H_5^+$

    Within experimental error, 3 moles of $N_2H_5^+$ react with 2 moles of $BrO_3^-$.

3.  The reduction of 2 moles of $BrO_3^-$ and oxidation of 3 moles of $N_2H_5^+$ involves 12 electrons.

    Each mole of $N_2H_5^+$ must undergo a 4 electron oxidation. Since $N_2H_5^+$ contains two N atoms, the oxidation number of each N atom must increase by 2. In $N_2H_5^+$ the oxidation number of N is -2 so that in the product, the oxidation number is zero. The product must consist of $N_2$ ie the colourless gas. Hence the equation for the reaction is as follows:

    $$2BrO_3^- + 3N_2H_5^+ \rightarrow 2Br^- + 3N_2 + 3H^+ + 6H_2O$$

**Acknowledgement**
Dr Eddie Allen, University of Portsmouth

# Calculations with sodium sulfate

**Problem**

1. If 10 g of $Na_2SO_4$ is dissolved in water and made up to a total volume of 500 $cm^3$, calculate the concentration of the solution in mol $dm^{-3}$.

2. What amount of sodium ion, $Na^+$, would be in 1 $dm^3$ of the solution in part 1 above?

3. If a class of 150 students are each to be supplied with 100 $cm^3$ of 0.10 mol $dm^{-3}$ $Na_2SO_4$ solution, calculate the total volume of solution required, and the mass of $Na_2SO_4$ needed.

4. If 10 g of sodium sulfate decahydrate, $Na_2SO_4.10H_2O$, rather than the anhydrous salt used in part 1 above, is dissolved in 500 $cm^3$ water, calculate the concentration of the sodium sulfate solution obtained.

**Prior knowledge**
Calculations.

**This problem is suitable for:**
■ First year students
■ A tutorial – group work (2–3 students)
■ Approximately 30 minutes

**Knowledge/skills gained**
Revision of basic principles.

## T4

# Calculations with sodium sulfate

**Solution**

1.  Molar mass of $Na_2SO_4$ is 142 g $mol^{-1}$

    Amount   = mass/molar mass
                   = 10 g/142 g $mol^{-1}$
                   = 0.07 mol, in 500 $cm^3$

    Amount   = concentration x (volume/1000) $dm^3$
    0.07 mol   = concentration x (500/1000) $dm^3$
    Therefore, concentration = 0.14 mol $dm^{-3}$

2.  $Na_2SO_4 \rightarrow 2Na^+_{(aq)} + SO_4^{2-}_{(aq)}$
    *ie* amount of $Na^+$ = 2 x amount of $Na_2SO_4$ = 0.28 mol

3.  Each student requires 100 $cm^3$ of 0.10 mol $dm^{-3}$ $Na_2SO_4$ solution
    Therefore, total volume required is 150 x 0.1 $dm^3$ = 15 $dm^3$

    Amount of $Na_2SO_4$ needed = 15 $dm^3$ x 0.1 mol $dm^{-3}$ = 1.5 mol
    Therefore, mass of $Na_2SO_4$ required is 1.5 mol x 142 g $mol^{-1}$ = 213 g

4.  Molar mass of $Na_2SO_4.10H_2O$ is 322.2 g $mol^{-1}$

    Amount of decahydrate   = 10 g/322.2 g $mol^{-1}$
                                  = 3.11 x $10^{-2}$ mol, in 500 $cm^3$

    Amount = concentration x (volume/1000) $dm^3$
    3.11 x $10^{-2}$ mol = concentration x (500/1000) $dm^3$
    Therefore, concentration of the sodium sulfate solution = 0.06 mol $dm^{-3}$

# A calculation involving determination of nickel using dimethylglyoxime

**Problem**

What volume of 2.15% w/w dimethylglyoxime in alcohol is needed to provide an excess of 50% for the determination of nickel in 0.99 g of steel containing 2.07% w/w nickel?

Assume that the density of the dimethylglyoxime solution is 0.79 g cm$^{-3}$. The reaction is:

$$\text{Ni}^{2+} \quad + \quad 2\text{Hdmg} \quad \rightarrow \quad \text{Ni(dmg)}_2 \quad + \quad 2\text{H}^+$$

| Ni$^{2+}$ | 2Hdmg | Ni(dmg)$_2$ | |
|---|---|---|---|
| 58.71 | 116.12 | 288.93 | Molar mass (g mol$^{-1}$) |

where    dmg   = dimethylglyoxime anion
          Hdmg = dimethylglyoxime

**Prior knowledge**
Calculations.

**This problem is suitable for:**
■ First year students
■ A tutorial – group work (2–3 students)
■ Approximately 20 minutes

**Knowledge/skills gained**
Calculation skills.

# A calculation involving determination of nickel using dimethylglyoxime

This problem can be solved in groups of 2–3 students. A lot of the information necessary for the calculation has been given in the question.

**Solution**

Mass of Ni in sample
$= (2.07/100) \times 0.99$ g
$= 0.0205$ g

Amount of Ni in sample
$= 0.0205/58.71 = 3.49 \times 10^{-4}$ mol

1 mole of $Ni^{2+}$ reacts with 2 moles of Hdmg
Therefore, amount of Hdmg
$= 2 \times 3.49 \times 10^{-4}$ mol $= 6.98 \times 10^{-4}$ mol

Require 50% excess of Hdmg
Therefore, amount of Hdmg required
$= 6.98 \times 10^{-4} \times 1.5$ mol
$= 1.05 \times 10^{-3}$ mol

mass of Hdmg
$= 1.05 \times 10^{-3}$ mol $\times 116.12$ g mol$^{-1}$
$= 0.1219$ g

Have 2.15% w/w solution of Hdmg
Therefore, mass of solution required $= (0.1219/2.15) \times 100 = 5.67$ g

Finally, calculate volume of Hdmg required:

Density, $\rho$ = mass/volume
Therefore, volume
$= 5.67$ g$/0.79$ g cm$^{-3}$
$= 7.18$ cm$^3$

**Acknowledgement**
Dr John Smith, University of Newcastle upon Tyne

## S6

# Oxidation state and stoichiometric equations

**Problem**

1. Work out the oxidation state of each sulfur atom in the following species:

$$SO_4^{2-}, \quad SO_3^{2-}, \quad S_2O_3^{2-}, \quad S_2O_4^{2-}, \quad S_2O_6^{2-}, \quad S_2O_8^{2-}$$

2. Using the above sulfur species, and water and hydrogen ions where necessary, complete equations (I) and (II) by working out what A, B, C, D, E and F are:

   (I)    $2Cr^{3+} + 3S_2O_8^{2-} + A \rightarrow Cr_2O_7^{2-} + B + C$

   (II)    $2S_2O_4^{2-} + D \rightarrow 2SO_3^{2-} + E + F$

**Prior knowledge**
Oxidation states; basic concepts of redox reactions.

**This problem is suitable for:**
■ First year students
■ A tutorial – group work (2–3 students)
■ Approximately 1 hour

**Knowledge/skills gained**
Stoichiometric equations.

T6

# Oxidation state and stoichiometric equations

This problem can be used to teach students the techniques involved in solving redox reactions. The students may need a brief summary of redox reactions.

The following knowledge is required:

(i) Oxidation involves the loss of electrons or increase in oxidation number.

(ii) Conversely reduction involves electron gain or decrease in oxidation number.

(iii) Disproportionation is a self oxidation and reduction process.

(iv) In a redox reaction the total number of electrons released in the oxidation must equal the total number of electrons gained in the reduction.

(v) A combination of the appropriate half equation for reduction with that for oxidation will give the required equation for the reaction occurring.

**Solution**

1. Denote the oxidation number of S as $\underline{\underline{S}}$

For $SO_4^{2-}$,

$$\underline{\underline{S}} + 4(-2) = -2$$

Therefore,     $\underline{\underline{S}} = +6$

|  | $SO_4^{2-}$ | $SO_3^{2-}$ | $S_2O_3^{2-}$ | $S_2O_4^{2-}$ | $S_2O_6^{2-}$ | $S_2O_8^{2-}$ |
|---|---|---|---|---|---|---|
| Oxidation Number | +6 | +4 | +2 | +3 | +5 | +7 |

2. (I)     $2Cr^{3+} + 3S_2O_8^{2-} + A \rightarrow Cr_2O_7^{2-} + B + C$

3 moles of $S_2O_8^{2-}$ must undergo a total change in oxidation number of 6
*ie*     $3S_2O_8^{2-} + 6e \rightarrow 6SO_4^{2-}$
$\underline{\underline{S}} = +7 \qquad\qquad \underline{\underline{S}} = +6$

Therefore,     $2Cr^{3+} + 3S_2O_8^{2-} + A \rightarrow Cr_2O_7^{2-} + 6SO_4^{2-} + C$

Add water to balance equation, giving

$$2Cr^{3+} + 3S_2O_8^{2-} + 7H_2O \rightarrow Cr_2O_7^{2-} + 6SO_4^{2-} + 14H^+$$

(II)    $2S_2O_4^{2-} + D \rightarrow 2SO_3^{2-} + E + F$
        $\underline{\underline{S}} = +3 \qquad\qquad \underline{\underline{S}} = +4$

Two moles of $SO_3^{2-}$ must be derived from 1 mole of $S_2O_4^{2-}$ with a total oxidation state change of 2. Therefore, in this disproportionation 1 mole of $S_2O_4^{2-}$ must also undergo an oxidation state change of 2 to give the reduced product.

$$S_2O_4^{2-} + 2H_2O \rightarrow 2SO_3^{2-} + 4H^+ + 2e$$

From the stoichiometry there must be a further reaction involving two electrons to give another sulfur species.

$S_2O_4^{2-} + 2H^+ + 2e \rightarrow S_2O_3^{2-} + H_2O$
$\underline{\underline{S}} = +3 \qquad\qquad\qquad \underline{\underline{S}} = +2$

The overall reaction is:

$$2S_2O_4^{2-} + H_2O \rightarrow 2SO_3^{2-} + S_2O_3^{2-} + 2H^+$$

**Acknowledgement**

Dr Eddie Allen, University of Portsmouth

S7

# Investigating reactions between copper(II) and hydrazine

**Problem**

In an investigation of the reaction between copper(II) ions, $Cu^{2+}$, and hydrazine, $N_2H_4$, in alkaline conditions the following observations were made:

25 $cm^3$ of a solution containing 25 g $dm^{-3}$ $CuSO_4.5H_2O$ was mixed with aqueous sodium hydroxide. Then 25 $cm^3$ of a solution containing 13 g $dm^{-3}$ $N_2H_6SO_4$ was added. After the evolution of an inert gas had ceased, a fine orange-red precipitate was observed. This precipitate was removed and the excess hydrazine was found to be equivalent to 75 $cm^3$ of $2.50 \times 10^{-2}$ mol $dm^{-3}$ acidified potassium iodate(V), when the following reaction occurred:

$$IO_3^- + N_2H_4 + 2H^+ \rightarrow N_2 + I^+ + 3H_2O$$

Deduce the equation for the reaction between copper(II) ions and hydrazine in alkaline conditions.

Relative atomic masses:
Cu = 63.54, S = 32.07, O = 16.01, N = 14.00, H = 1.01

**Prior knowledge**
Mole calculations.

**This problem is suitable for:**
■ Second year students
■ A tutorial – group work (2–3 students)
■ Approximately 45 minutes

**Knowledge/skills gained**
Balancing equations.

# Investigating reactions between copper(II) and hydrazine

The suggested approach for this problem is small group work. Students, in discussion within their groups, will probably come up with some proposals for the products of the reaction, thereby suggesting the resulting equation. These deductions will be helpful to the students in their calculations.

Students may deduce that the gas produced is nitrogen and that the red precipitate is probably $Cu_2O$. These deductions are obviously correct and discussion between group members to achieve this is intended. Students should then be encouraged to give reasons behind their suggestions.

## Solution

Molar mass of $CuSO_4.5H_2O$ is 249.70 g mol$^{-1}$
Molar mass of $N_2H_6SO_4$ is 130.17 g mol$^{-1}$

$[CuSO_4.5H_2O]$ = 25 g dm$^{-3}$/249.70 g mol$^{-1}$ = 0.10 mol dm$^{-3}$
$[N_2H_6SO_4]$ = 13 g dm$^{-3}$/130.17 g mol$^{-1}$ = 0.10 mol dm$^{-3}$

$$IO_3^- + N_2H_4 + 2H^+ \rightarrow N_2 + I^+ + 3H_2O$$

From the balanced equation, $N_2H_4$ and $IO_3^-$ react in a 1:1 mole ratio.

Excess $N_2H_4$ = 2.50 x 10$^{-2}$ mol dm$^{-3}$ x (75/1000) dm$^3$ = 1.88 x 10$^{-3}$ mol, from titration with acidified potassium iodate(V).

Calculate amount of $N_2H_4$ reacted with $Cu^{2+}$:
[0.10 mol dm$^{-3}$ x (25/1000) dm$^3$] - 1.88 x 10$^{-3}$ mol = 0.62 x 10$^{-3}$ mol

*ie* 0.62 x 10$^{-3}$ mol $N_2H_4$ reacts with 0.10 mol dm$^{-3}$ x (25/1000) dm$^3$ $Cu^{2+}$

$\Rightarrow$ 1 mole $N_2H_4$ reacts with
[0.10 mol dm$^{-3}$ x (25/1000) dm$^3$/0.62 x 10$^{-3}$ mol] $Cu^{2+}$ ≡ 4 mol $Cu^{2+}$

Hydrazine is oxidised to give nitrogen gas.

In alkaline conditions:
$$N_2H_4 + 4OH^- \rightarrow N_2 + 4H_2O + 4e$$

Since 1 mole of $N_2H_4$ reacts with 4 moles of $Cu^{2+}$, each mole of $Cu^{2+}$ undergoes a one-electron reduction.
Therefore, the orange-red precipitate is $Cu_2O$ *ie* Cu(I).

$$4Cu^{2+} + 4OH^- + 4e \rightarrow 2Cu_2O + 2H_2O$$

The overall reaction is:

$$4Cu^{2+} + N_2H_4 + 8OH^- \rightarrow 2Cu_2O + N_2 + 6H_2O$$

**Acknowledgement**
Dr Eddie Allen, University of Portsmouth

RS•C

# Stoichiometry – the reaction of iron(III) with hydroxylammonium chloride

**Problem**

A solution containing 0.14 g of hydroxylammonium chloride, $(NH_3OH)Cl$, in sulfuric acid was heated with a solution of 0.10 mol dm$^{-3}$ ammonium iron(III) sulfate, $(NH_4)Fe(SO_4)_2.12H_2O$.

A colourless gas was evolved and the Fe(III) was partly reduced to Fe(II).

After cooling, phosphoric(V) acid was added and the solution was titrated against potassium dichromate solution. The mean titre was 37.60 cm$^3$.

A 25.00 cm$^3$ aliquot of the 0.10 mol dm$^{-3}$ solution of the iron(III) salt was reduced with tin(II) chloride solution. This was then treated with a slight excess of mercury(II) chloride solution. The resulting iron(II) solution was titrated against the same solution of potassium dichromate. The titre was 23.80 cm$^3$.

1.  Calculate the concentration of the potassium dichromate solution.

2.  Calculate the number of moles of Fe$^{3+}$ which react with one mole of the hydroxylammonium chloride.

3.  Write a balanced equation for the reaction of the hydroxylammonium cation with Fe$^{3+}$ under acidic conditions and hence identify the colourless gas evolved.

Molar mass of $(NH_3OH)Cl$ is 69.49 g mol$^{-1}$.

**Prior knowledge**
Balancing equations; interpretation of data.

**This problem is suitable for:**
∎ Second year students
∎ A tutorial – group work (2–3 students)
∎ Approximately 45 minutes

**Knowledge gained**
Redox reactions; calculations of molarity.

**T8**

# Stoichiometry – the reaction of iron(III) with hydroxylammonium chloride

**Solution**

1. Using the part of the problem involving a 25 cm$^3$ sample of the iron(III)
   solution, which has been reduced to iron(II) by tin(II). Potassium dichromate
   is an oxidising agent, therefore the iron is oxidised to its +3 state. Thus the
   following equation can be derived:

   $$6Fe^{2+} + Cr_2O_7^{2-} + 14H^+ \rightarrow 6Fe^{3+} + 2Cr^{3+} + 7H_2O$$

   25 cm$^3$ of 0.10 mol dm$^{-3}$ iron(II) solution was used.
   23.80 cm$^3$ of the potassium dichromate solution was needed.
   From the equation, 6 moles of Fe$^{2+}$ react with 1 mole of Cr$_2$O$_7^{2-}$.

   | | |
   |---|---|
   | Amount of Fe$^{2+}$ | = 0.10 mol dm$^{-3}$ x (25/1000) dm$^3$ |
   | | = 2.50 x 10$^{-3}$ mol |
   | Amount of Cr$_2$O$_7^{2-}$ | = 2.50 x 10$^{-3}$ mol/6 |
   | | = 4.17 x 10$^{-4}$ mol |
   | Amount | = concentration x volume |

   Therefore, concentration of potassium dichromate solution
   
   = 4.17 x 10$^{-4}$ mol/(23.80/1000) dm$^3$
   
   = 1.75 x 10$^{-2}$ mol dm$^{-3}$

2. The concentration of K$_2$Cr$_2$O$_7$ is 1.75 x 10$^{-2}$ mol dm$^{-3}$ (as calculated in part 1).
   Let the amount of potassium dichromate used to oxidise Fe(II) → Fe(III) be X.
   X = 1.75 x 10$^{-2}$ mol dm$^{-3}$ x (37.60/1000) dm$^3$ = 0.66 x 10$^{-3}$ mol

   | | |
   |---|---|
   | Ratio of Fe(II):Cr$_2$O$_7^{2-}$ | = 6:1 |
   | Ratio of Fe(II):Fe(III) | = 1:1 |

   Therefore, amount of Fe$^{2+}$ produced by reduction of Fe$^{3+}$
   
   = 0.66 x 10$^{-3}$ mol x 6 = 3.96 x 10$^{-3}$ mol

   | | |
   |---|---|
   | Amount of (NH$_3$OH)Cl | = mass/molar mass |
   | | = 0.14 g/69.49 g mol$^{-1}$ |
   | | = 2.01 x 10$^{-3}$ mol |

   The mole ratio Fe$^{3+}$ : NH$_3$OH$^+$ = 3.96 : 2.01 = 2:1
   *ie* 2 moles of Fe$^{3+}$ react with 1 mole of NH$_3$OH$^+$

3. Reduction of 2Fe$^{3+}$ → 2Fe$^{2+}$ involves a gain of 2 electrons, hence NH$_3$OH$^+$
   must lose two electrons.
   Hence  NH$_3$OH$^+$ → $\frac{1}{2}$N$_2$O + 3H$^+$ + $\frac{1}{2}$H$_2$O + 2e
   *ie* gas evolved is N$_2$O
   Balanced equation:
   $$2NH_3OH^+ + 4Fe^{3+} \rightarrow N_2O + 4Fe^{2+} + 6H^+ + H_2O$$

**Acknowledgement**
Dr Eddie Allen, University of Portsmouth

# Identifying a sulfate using gravimetric analysis

**Problem**

Sulfate ions and barium ions form a precipitate, when reacted.

1.58 g of a sulfate, Q, yielded 1.15 g of $BaSO_4$ on addition of excess aqueous $BaCl_2$.

A second sample of 2.74 g of Q lost 1.53 g of water when heated strongly until its mass was constant.

Q has the formula $X_mSO_4.pH_2O$; determine p and the relative atomic mass of $X_m$, and hence identify Q.

**Prior knowledge**
Calculations.

**This problem is suitable for:**
■ Second year students
■ A tutorial – group or individual work
■ Approximately 30 minutes

**Knowledge/skills gained**
Revision of mole equations.

# T9

# Identifying a sulfate using gravimetric analysis

This problem involves inserting the data in appropriate equations.

**Solution**

$$X_mSO_4.pH_2O + BaCl_2 \rightarrow BaSO_4 + 2XCl + pH_2O$$
1.58 g                 1.15 g

Molar mass of $BaSO_4$ is 233.39 g $mol^{-1}$
Molar mass of $SO_4^{2-}$ is 96.06 g $mol^{-1}$

Sulfate content:
233.39 g $mol^{-1}$ $BaSO_4$ contains 96.06 g $SO_4^{2-}$
Therefore, 1.15 g $BaSO_4$ contains 1.15 g x 96.06 g $mol^{-1}$/233.39 g $mol^{-1}$
$$= 0.47 \text{ g } SO_4^{2-}$$
Therefore, 1.58 g Q contains 0.47 g $SO_4^{2-}$

Water content:
2.74 g Q contains 1.53 g $H_2O$ *ie* (1.53/2.74) x 100% = 55.8% of Q is water.
Therefore, 1.58 g Q contains (1.58 x 55.8)/100 g $H_2O$ = 0.88 g

Metal (X) content:
1.58 g Q contains [1.58 g - (0.88 + 0.47)] g X       = 0.23 g

Calculate p:
Mole ratio of water:sulfate is 0.88 g/18 g $mol^{-1}$:0.47 g/96.06 g $mol^{-1}$
$= 4.89 \times 10^{-2}:4.89 \times 10^{-3} = 10:1$
$\Rightarrow p = 10$

Content:
sulfate:     0.47 g
water:       0.88 g
metal, X:   0.23 g

Formula of Q:

Let A                 = molar mass of X *ie* mA is the molar mass of $X_m$
0.47 g/96.06 g $mol^{-1}$ $SO_4^{2-}$ = 0.23 g/mA g $mol^{-1}$ X
$4.89 \times 10^{-3}$ mol $SO_4^{2-}$    = 0.23 g/mA g $mol^{-1}$ X
mA                  = 47.03 g $mol^{-1}$
From mass and valence considerations $X_m$ must be $Na_2$ (since Na has molar mass of 22.99 g $mol^{-1}$).

Therefore, Q is $Na_2SO_4.10H_2O$

**Acknowledgement**
Dr John Smith, University of Newcastle upon Tyne

S10

# Determining the molecular weight and $K_a$ for a weak acid

**Problem**

A weak acid, X, is known to be monoprotic and crystalline at room temperature.

Using the following information calculate the molar mass of X and the value of its acidity constant, $K_a$.

A small quantity (0.50 g) of X was dissolved in 50 $cm^3$ of water. This solution was titrated with 0.10 mol $dm^{-3}$ sodium hydroxide. The results from the titration are given below:

| $cm^3$ of NaOH | 0.00 | 9.00 | 20.05 | 22.00 | 31.00 | 41.05 (end point) |
|---|---|---|---|---|---|---|
| pH | 2.65 | 3.05 | 6.00 | 7.90 | 8.35 | 8.48 |

**Prior knowledge**
Plotting titration curves; interpolating data to perform calculations; acid-base titrations.

**This problem is suitable for:**
■  First year students
■  A tutorial – group or individual work
■  Approximately 30 minutes

**Knowledge/skills gained**
The theory behind titrations.

# T10

# Determining the molecular weight and $K_a$ for a weak acid

Students should know that the end point of a reaction is also known as the equivalence point *ie* neutralisation.

**Solution**

A graph of pH versus Volume of NaOH added should be drawn.

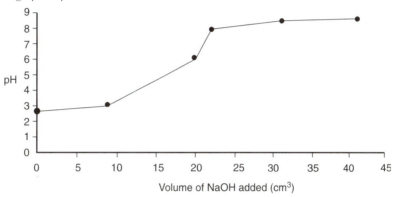

Calculate molar mass, X, of acid:

amount = mass/molar mass
amount = concentration x volume

At the equivalence point:

amount of acid = amount of base
amount of NaOH = 0.10 mol dm$^{-3}$ x (41.05/1000) dm$^3$
amount of acid = 0.50 g/X

Therefore,     0.50 g/X = 0.10 mol dm$^{-3}$ x (41.05/1000) dm$^3$
0.50 g/X = 4.105 x 10$^{-3}$ mol
Therefore,     X = 121.8 g mol$^{-1}$
*ie* molar mass of acid = 121.8 g mol$^{-1}$

Calculate $K_a$:

There is a time during a weak acid–strong base titration when both the acid and base are equal in concentration. This is when half the base has been added and half the acid has been neutralised.

*ie*     41.05 cm$^3$/2 = 20.53 cm$^3$
The pH after addition of 20.53 cm$^3$ is 6.47, from graph.
At this stage, pH = p$K_a$
Therefore, $K_a$ = log$^{-1}$ (p$K_a$) = 3.39 x 10$^{-7}$ mol dm$^{-3}$

## S11

# Determining chloride in a mixture of ferrous chloride and ferrous ammonium sulfate

**Problem**

A solid mixture weighing 0.55 g contained only ammonium iron(II) sulfate and iron(II) chloride.

The sample was dissolved in 1.00 mol dm$^{-3}$ $H_2SO_4$, oxidised to Fe(III) with $H_2O_2$, and precipitated with cupferron. The iron(III)-cupferron complex was heated strongly to produce 0.17 g of $Fe_2O_3$. Calculate the percentage by mass of Cl in the original sample.

| $FeSO_4.(NH_4)_2SO_4.6H_2O$ | $FeCl_2.6H_2O$ | $C_6H_5N(NO)ONH_4^+$ |
|---|---|---|
| Ammonium iron(II) sulfate | Iron(II) chloride | Cupferron |
| Molar mass (g mol$^{-1}$) | | |
| 392.13 | 234.84 | 155.16 |

**Prior knowledge**
Solving equations.

**This problem is suitable for:**
■  First year students
■  A tutorial – group or individual work
■  Approximately 30 minutes

**Knowledge/skills gained**
Calculations.

T11

# Determining chloride in a mixture of ferrous chloride and ferrous ammonium sulfate

This problem could be set as individual work and discussed in a group tutorial.

**Solution**

0.17 g of $Fe_2O_3$ contains $(112/159.69) \times 0.17$ g of Fe
$\qquad\qquad\qquad\qquad = 0.119$ g of Fe
amount of Fe in $Fe_2O_3 \qquad = 0.119$ g/56 g $mol^{-1}$
$\qquad\qquad\qquad\qquad = 2.13 \times 10^{-3}$ mol
Therefore, $2.13 \times 10^{-3}$ mol of Fe in solid mixture.

Let X $\qquad\qquad\qquad$ = amount of iron(II) chloride
Therefore, $(2.13 \times 10^{-3} - X)$ = amount of ammonium iron(II) sulfate.

Hence,
$(2.13 \times 10^{-3} - X)\ 392.13 + 234.84X = 0.55$ g
$\Rightarrow \qquad X \qquad\qquad = 1.81 \times 10^{-3}$ mol
mass of Cl $\qquad\qquad = 2 \times 35.45$ g $mol^{-1} \times 1.81 \times 10^{-3}$ mol
$\qquad\qquad\qquad\qquad = 0.129$ g
% by mass of Cl $\qquad = (0.129/0.55) \times 100\%$
$\qquad\qquad\qquad\qquad = 23.4\%$

**Acknowledgement**
Dr John Smith, University of Newcastle upon Tyne

# Energy changes – hot air balloons

**Problem**

Bristol is famous for the manufacture of hot air balloons. Each August there is a balloon festival and some 300 balloons are launched. The mass of a spherical balloon, basket and propane cylinders is typically 400 kg, and the diameter of a balloon is about 20 m. If, on a frosty morning at 0 °C the air in a balloon is heated to 60 °C, what load can be lifted in the basket?

Discuss any assumptions you feel should be made before beginning the calculation.

**Prior knowledge**
Physical chemistry calculations.

**This problem is suitable for:**
■  Second year students
■  A tutorial – discussion followed by group or individual work
■  Approximately 45 minutes

**Knowledge/skills gained**
This problem requires a degree of preparation before it can be solved. This technique should teach students that a problem requires careful consideration before it is solved.

# T12

# Energy changes – hot air balloons

Not a lot of information has been given in this problem. Some assumptions should be discussed and agreed before starting the calculation.

The following assumptions should be made:
- barometric pressure at ground level is 101.33 kPa;
- air consists of 20% oxygen and 80% nitrogen (by mass);
- no gas is lost from the balloon; and
- the balloon does not expand as the temperature, or the balloon, rises.

**Note:** the lift due to the balloon is given by the difference in density between the warm and cool air multiplied by the volume of the balloon. This part will require some discussion as it may not be obvious or recalled by all students.

## Solution

Firstly calculate the mass of 1 $m^3$ of air at 0 °C.
Then calculate the mass of 1 $m^3$ of air at 60 °C.

Molar mass of $O_2$ is 32 g $mol^{-1}$
Molar mass of $N_2$ is 28 g $mol^{-1}$

Air contains 20% oxygen and 80% nitrogen, by mass.
Therefore, effective molar mass of air
$$= \{(80/100) \times 28 \text{ g mol}^{-1} + (20/100) \times 32 \text{ g mol}^{-1}\}$$
$$= 28.8 \text{ g mol}^{-1}$$

| | |
|---|---|
| Density of 1 $m^3$ of air at 0 °C | = 1.29 kg $m^{-3}$ |
| Density of 1 $m^3$ of air at 60 °C | = 1.05 kg $m^{-3}$ |

Calculate the volume of the balloon:

| | |
|---|---|
| Volume of sphere | $= (4/3)\pi r^3$ |
| Volume of balloon | $= (4/3)\pi(10\text{m})^3$ |
| | $= 4188.80 \text{ m}^3$ |
| Load | $= \{4188.80 \text{ m}^3 \times (1.29 \text{ kg m}^{-3} - 1.05 \text{ kg m}^{-3})\}$ |
| | $= 1005.31 \text{ kg}$ |

The total mass of the balloon, basket and propane cylinders is 400 kg (from problem)

| | |
|---|---|
| Therefore, load that can be lifted | = 1005.31 kg - 400 kg |
| | = 605.31 kg. |

**Acknowledgement**
Dr C J Wormald, University of Bristol

S13

# Stalagmites and stalactites

## Problem

Stalagmites and stalactites are columns of calcium carbonate that can grow from the ground and the ceiling under bridges on motorways, or in caves for example.

Water saturated with calcium hydrogen carbonate drips down and as the water evaporates it leaves behind calcium carbonate. For this reason, stalagmites and stalactites continuously increase in size.

Estimate the number of drops of water that fall on to a stalagmite every day. Treat the stalagmite as a cylinder of diameter 0.3 cm that grows in length by 2 cm every 100 years with the width remaining constant.

In any calculations you perform assume that the water is saturated with calcium carbonate and that the drops are spheres with diameter 3 mm. Assume, also, that the deposition is entirely from saturated aqueous calcium carbonate.

### Prior knowledge
Calculations; solubility products.

### This problem is suitable for:
- First year students
- A tutorial – group work
- Approximately 30 minutes

### Knowledge/skills gained
Improved ability in calculations; retrieving information.

# T13

# Stalagmites and stalactites

A suggested approach for this problem is given below. The solubility product of calcium carbonate is required – encourage students to find it.

**Solution**

At 25 °C $K_s(CaCO_3) = 8.70 \times 10^{-9}$ mol$^2$ dm$^{-6}$

As there are no other ions present then $CaCO_3 \rightarrow Ca^{2+}_{(aq)} + CO_3^{2-}_{(aq)}$
and $[Ca^{2+}] = [CO_3^{2-}]$

Therefore, solubility of $[Ca^{2+}]$   = $(8.70 \times 10^{-9}$ mol$^2$ dm$^{-6})^{0.5}$
                                         = $9.33 \times 10^{-5}$ mol dm$^{-3}$

For the purpose of the problem we will assume the drops to be spheres of 3 mm diameter (*ie* radius 0.15 cm).

Volume of a sphere                = $4/3\pi r^3$
                                    = $4/3 \times \pi \times (0.15$ cm$)^3$
                                    = $1.41 \times 10^{-2}$ cm$^3$

The amount of $CaCO_3$ in each raindrop can now be calculated:
                    = $(9.33 \times 10^{-5}$ mol dm$^{-3}) \times (1.41 \times 10^{-2}$ cm$^3$)/1000
                    = $1.32 \times 10^{-9}$ mol

As volume         = mass/density
                    = $(1.32 \times 10^{-9}$ mol) $\times 100.10$ g mol$^{-1}$/2.70 g cm$^{-3}$
                    = $4.88 \times 10^{-8}$ cm$^3$

The total volume of stalagmite formed in 100 years
                    = $\pi r^2 l$
                    = $\pi \times (0.15$ cm$)^2 \times 2$ cm
                    = $0.14$ cm$^3$

The number of drops necessary to give this volume
                    = $0.14$ cm$^3$/$4.88 \times 10^{-8}$ cm$^3$
                    = $2.87 \times 10^6$ drops

Number of drops/100 years        = $2.87 \times 10^6$

Therefore number of drops per day
                          = $2.87 \times 10^6$/365 $\times$ 100
                          = $78.6$
                          $\cong$ 79 drops per day

S14

# Enthalpy of formation of krypton(II) fluoride

**Problem**

The bond enthalpy for the Kr-F bond can be estimated using the equation:

$$B_{(Kr-F)} = \sqrt{96.50(X_F - X_{Kr})^2 + (B_{(Kr-Kr)} \times B_{(F-F)})} \quad kJ\ mol^{-1}$$

Where X is the electronegativity of the given atom and B the bond enthalpy for the indicated bond.

In one such calculation estimated values of $X_{Kr} = 3.5$ and $B_{(Kr-Kr)} = 1\ kJ\ mol^{-1}$ were used along with the known values of $X_F = 4$ and $B_{(F-F)} = 158\ kJ\ mol^{-1}$.

Discuss the values of B and X estimated for krypton and use them to estimate the enthalpy of formation of $KrF_{2\ (g)}$.

$KrF_2$ can be synthesised and is a thermally unstable white solid. In view of the results of the calculation above discuss the synthesis.

**Prior knowledge**
Bond enthalpy; enthalpies of sublimation and formation.

**This problem is suitable for:**
■  Second year students
■  A tutorial – group or individual work
■  Approximately 45 minutes

**Knowledge/skills gained**
Analysing data.

## T14

# Enthalpy of formation of krypton(II) fluoride

**Solution**

$B_{(Kr-F)} = 11.53$ kJ mol$^{-1}$ from the data

Using this information in the equation below gives the enthalpy of formation

$Kr_{(g)} + F_{2(g)} \rightarrow KrF_{2(g)}$

as $+134.9$ kJ mol$^{-1}$ *ie* endothermic.

The low estimate for $B_{(Kr-Kr)}$ seems reasonable as $Kr_2$ does not exist and Kr has a low boiling point. The high value of $X_{Kr}$ may seem strange but is reasonable when the definition of electronegativity is remembered and given the position of Kr in the periodic table.

The enthalpy change for the reaction is greater than zero (*ie* positive) and so the reaction, as written, seems unlikely.

However, as $KrF_2$ is a solid, the process for formation is correctly written as

$Kr_{(g)} + F_{2(g)} \rightarrow KrF_{2(g)}$ and so the enthalpy change for $KrF_{2(g)} \rightarrow KrF_{2(s)}$ (enthalpy of sublimation) must be taken into account in the enthalpy of formation and this will then favour the forward reaction.

The entropy change is less than zero (*ie* negative) (2 moles of gas giving 1 mole of solid) adding an unfavourable contribution to the change in free energy, $\Delta G$.

Values of $B_{(Kr-F)}$ can only be approximate and the idea is to discuss the high value of electronegativity. For instance in any bond to Kr where the Kr is sharing electrons there would be a strong tendency for it to attract electrons to itself (*ie* it would be highly electronegative).

For the synthesis from atomic fluorine the enthalpy change is simply $-2B_{(Kr-F)}$ and so is bound to be exothermic. It should be noted that this is not the enthalpy of formation and that decomposition back to Kr and $F_2$ is still favourable.

**Acknowledgement**
Dr Andy Platt, Staffordshire University

## S15

# Deriving balanced equations for reactions of xenon tetrafluoride and the xenate ion

### Problem

1. (a) Xenon tetrafluoride reacts vigorously with water giving xenon trioxide and xenon, as the only and final xenon-containing products in what appears to be a hydrolysis/disproportionation reaction.
Derive a balanced equation for this reaction.

   (b) In the early part of the above reaction, the observed ratio of $XeO_3$:Xe is 1:2 and oxygen is also produced. Suggest a reaction pathway that is consistent with these findings.

2. (a) The xenate ion, $(XeO_3(OH))^-$, slowly decomposes in alkaline solution to give perxenate ($XeO_6^{4-}$) and xenon as the only xenon-containing products.
Derive a balanced equation for this reaction.

   (b) It is observed that after the reaction in 2(a) 50% of the xenon is present as $XeO_6^{4-}$ and 50% as xenon, and that oxygen is evolved. Suggest a reaction pathway that explains this.

### Prior knowledge
Basic knowledge is needed for devising and balancing equations.

### This problem is suitable for:
- Second year students
- A tutorial – group work
- Approximately 45 minutes

### Knowledge/skills gained
Practice with deriving and balancing equations.

# T15

# Deriving balanced equations for reactions of xenon tetrafluoride and the xenate ion

**Hint:** $XeO_3$ is the most stable xenon oxide. $XeO_4$ exists and whilst other oxides are unknown, their presence as unstable intermediates is possible.

**Solution**

$$XeF_4 + H_2O \rightarrow XeO_3 + Xe + HF$$

Oxidation number          $+4$          $+6$     $0$

1. (a)  Balance equation
$$3XeF_4 + 6H_2O \rightarrow 2XeO_3 + Xe + 12HF$$

 (b)  Taking the equation in part 1 (a) students should be able to analyse the products and conclude that:
$$6XeF_4 + 12H_2O \rightarrow 4\ "XeO" +\quad 2XeO_4 + 24HF$$
$$\downarrow\qquad\qquad\downarrow$$
$$4Xe + 2O_2\quad 2XeO_3 + O_2$$
Therefore,
$$6XeF_4 + 12H_2O \rightarrow 4Xe + 2XeO_3 + 3O_2 + 24HF$$
according to the ratio given.

2. (a)  Predicted equation for the simple disproportionation:
$$4(XeO_3(OH))^- + 8OH^- \rightarrow 3XeO_6^{4-} + Xe + 6H_2O$$

 (b)  The reaction pathway is similar to that in 1 above but, now the initial disproportionation is to $XeO_6^{4-}$ and $XeO_2$, which subsequently decomposes to Xe and $O_2$ giving the overall equation:
$$2(XeO_3(OH))^- + 2OH^- \rightarrow XeO_6^{4-} + Xe + O_2 + 2H_2O$$

**Acknowledgement**
Dr Andy Platt, Staffordshire University

## S16

# Predicting formulae of metal carbonyls

**Problem**

For the complexes listed devise formulae for different arrangements of metal atoms and discuss possible analytical techniques by which the actual arrangement might be identified.

$Co_4(CO)_x$ ; $Mn_4(CO)_y$ ; and $Fe_3(CO)_z$.

Suggest structures for the formulae obtained.

### Prior knowledge
The 18-electron rule.

### This problem is suitable for:
■ Second year students
■ A tutorial – group work
■ Approximately 45 minutes

### Knowledge/skills gained
Ability to arrange atoms and to apply the 18-electron rule.

## T16

# Predicting formulae of metal carbonyls

The answer requires application of the 18-electron rule and the arrangement of metal atoms.

**Solution**

**$Co_4(CO)_x$**

For $Co_4$ linear, square and tetrahedral arrangements are possible, in principle, and each leads to a different molecular formula.

$Co_4$ requires 4 x 18 = 72 electrons in total

**$Co_4$ $d^9$ tetrahedral**

| | | |
|---|---|---|
| 4 x Co (0) | 4 x $d^9$ | 36e |
| 6 x Co–Co | | 12e |
| | | —— |
| | | 48e |
| | | —— |

CO provides 72 - 48 = 24e *ie* 12 CO
Predicted formula $Co_4(CO)_{12}$

**$Co_4$ $d^9$ square planar**

| | | |
|---|---|---|
| 4 x Co (0) | 4 x $d^9$ | 36e |
| 4 x Co–Co | | 8e |
| | | —— |
| | | 44e |
| | | —— |

CO provides 72 - 44 = 28e *ie* 14 CO
Predicted formula $Co_4(CO)_{14}$

**$Co_4$ $d^9$ linear**

| | | |
|---|---|---|
| 4 x Co (0) | 4 x $d^9$ | 36e |
| 3 x Co–Co | | 6e |
| | | —— |
| | | 42e |
| | | —— |

CO provides 72 - 42 = 30e *ie* 15 CO
Predicted formula $Co_4(CO)_{15}$

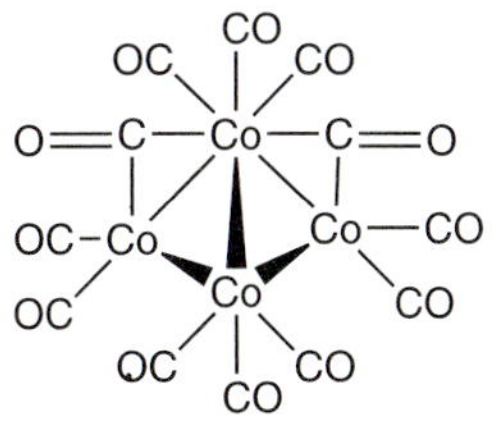

Suggested formula $Co_4(CO)_{12}$

**$Mn_4(CO)_y$**

$Mn_4$ requires $4 \times 18 = 72$ electrons in total.

**$Mn_4\ d^7$ tetrahedral**

| | |
|---|---|
| $4 \times Mn\ (0)\ 4 \times d^7$ | 28e |
| $6 \times Mn–Mn$ | 12e |
| | 40e |

CO provides $72 - 40 = 32e$ *ie* 16 CO
Predicted formula $Mn_4(CO)_{16}$

**$Mn_4\ d^7$ linear**

| | |
|---|---|
| $4 \times Mn\ (0)\ 4 \times d^7$ | 28e |
| $3 \times Mn–Mn$ | 6e |
| | 34e |

CO provides $72 - 34 = 38e$ *ie* 19 CO
Predicted formula $Mn_4(CO)_{19}$

**$Mn_4\ d^7$ square planar**

| | |
|---|---|
| $4 \times Mn\ (0)\ \ 4 \times d^7$ | 28e |
| $4 \times Mn–Mn$ | 8e |
| | 36e |

CO provides $72 - 36 = 36e$ *ie* 18 CO
Predicted formula $Mn_4(CO)_{18}$

For the square planar and linear arrangements the presence of bridging
carbonyls is required to maintain the 18-electron count around each metal.
This could be determined by infrared spectroscopy and related to the
bonding mechanism between CO and metals.

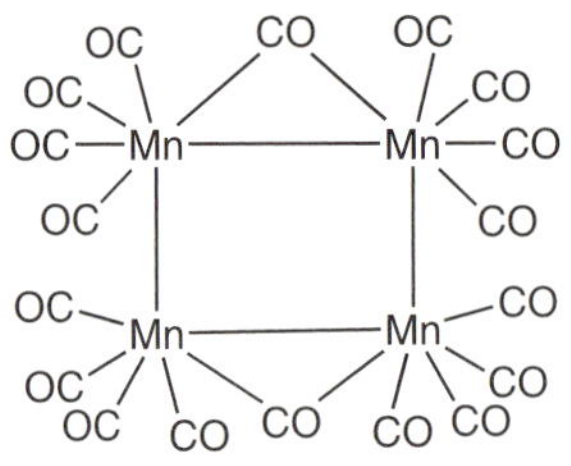

Suggested formula $Mn_4(CO)_{18}$

**Fe$_3$(CO)$_z$**
Fe$_3$ requires 3 x 18 = 54 electrons in total

| | |
|---|---|
| 3 x Fe (0)  3 x d$^8$ | 24e |
| 3 x Fe–Fe | 6e |
| | 30e |

CO provides 54 - 30 = 24e *ie* 12 CO
Predicted and suggested formula Fe$_3$(CO)$_{12}$

**Acknowledgement**

Dr Andy Platt, Staffordshire University

## S17

# Using mass spectrometry to distinguish isomers and identify compounds

**Problem**

1. How can mass spectrometry be used to distinguish between the following pair of isomers:
   hexan-2-one and 3-methylpentan-2-one.

2. A compound $C_4H_9X$ exhibited a molecular at ion at m/z = 136 and a second peak of equal intensity at m/z = 138 in the mass spectrum. Deduce what X is and briefly explain why two peaks are shown.

**Prior knowledge**
Isomerism and mass spectrometry.

**This problem is suitable for:**
■   First year students
■   A tutorial – group work
■   Approximately 30 minutes

**Knowledge/skills gained**
Occurrence of isotopes; application of mass spectrometry to distinguish isomers.

# T17

# Using mass spectrometry to distinguish isomers and identify compounds

## Solution

1. Students should initially draw both isomers. The fragmentation patterns should be discussed. Suggest the McLafferty rearrangement.

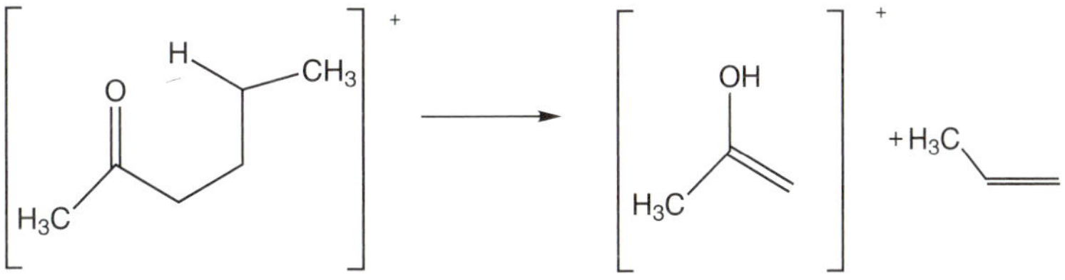

m/z = 58

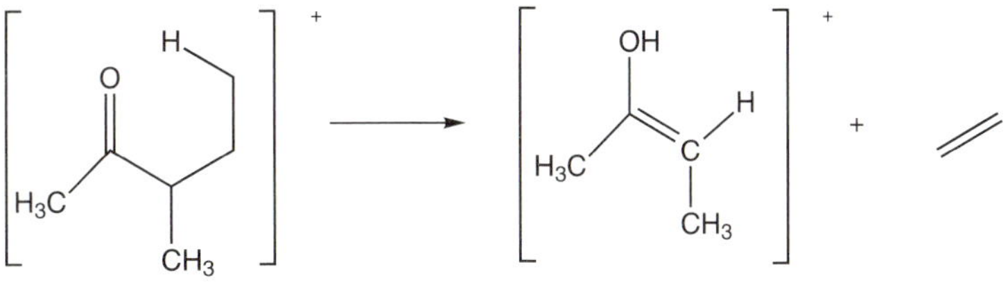

m/z = 72

A simple cleavage of an even-numbered molecular ion gives an odd-numbered fragment ion. However, this will not distinguish the isomers.

The isomers are likely to undergo a McLafferty rearrangement. To undergo such a rearrangement the molecule must possess an appropriately located heteroatom, a $\pi$ system and an abstractable hydrogen $\gamma$ to the C=O system.

The rearrangement and subsequent fragmentation of the isomers will result in distinguishable peaks in the mass spectrum.

2. Molecular ion at m/z = 136. The molecular formula is $C_4H_9X$.
   Mass of $C_4H_9 = 57$ g, therefore mass of X = 79 g.
   A second peak (M + 2) at m/z = 138.
   Mass of X = 81 g.

   Bromine exists as two isotopes with masses of 79 and 81 g and equal natural abundance. Therefore, the compound is $C_4H_9Br$.

**Acknowledgement**
Dr Tom Brown, University of Southampton

S18

# X-ray powder diffraction patterns for pentachloro-methylbenzene

**Problem**

Explain the following observation:

When the x-ray powder diffraction pattern for solid pentachloromethyl-benzene at room temperature was analysed, it indicated that electron density arising from the chloro and methyl groups is distributed uniformly around the ring.

### Prior knowledge
X-ray powder diffraction; rotator phase crystals.

### This problem is suitable for:
■ Second year students
■ Initial discussion followed by group or individual research leading to class discussion
■ Approximately 2 hours (including research time)

### Knowledge/skills gained
X-ray powder diffraction; rotator phase crystals.

# X-ray powder diffraction patterns for pentachloro-methylbenzene

The students may need to research this topic. This can be done individually or in groups but an ensuing class discussion is necessary to pool the results.

**Solution**

The observation is entirely inconsistent with the known molecular structure. It would be expected that the pattern should show differences between the chloro and methyl groups, as their electron densities are very different.

Pentachloromethylbenzene is a rotator phase crystal. The molecules continually jump over barriers due to reorientation (inter-molecular) and gain access to the six sites.

The molecular complex ions of the unit cell are continually changing due to the reorientations of all the molecules in the unit cell. This results in the x-ray pattern being an average of all these complex ions, hence the electron density appearing to be uniform around the ring.

**Acknowledgement**
Professor Graham Williams, University of Wales, Swansea

# Interpretation of the $^{19}$F NMR spectrum of $PF_3Cl_2$

**Problem**

Explain the following:

The $^{19}$F NMR spectrum of $PF_3Cl_2$ at ambient temperature shows a doublet. However, at -90 °C two signals are seen: a doublet of triplets (integrating for 1F) and a doublet of doublets (integrating for 2F).

### Prior knowledge
Isomerism; nuclear magnetic resonance spectroscopy.

### This problem is suitable for:
■ Second year students
■ Tutorial – small group work
■ Approximately 45–60 minutes

### Knowledge/skills gained
Analysis of spectra; valence shell electron pair repulsion theory (VSEPR).

## T19

# Interpretation of the $^{19}$F NMR spectrum of PF$_3$Cl$_2$

This problem can be solved by drawing structures of the compound PF$_3$Cl$_2$.

From the geometry of PF$_3$Cl$_2$ students should realise that it is trigonal bipyramidal. By discussing the structures within groups, students should come up with three possible isomers. Students may require assistance when assigning the signals to each structure and determining the fluorine atoms in chemically different environments.

Reference books should be available for details of the valence shell electron pair repulsion (VSEPR) theory and the interchanges of axial and equatorial sites (Berry Pseudorotations). The Berry mechanism applies to the trigonal bipyramidal-square planar-trigonal bipyramidal interconversion of five co-ordinate molecules.

**Solution**

$^{31}$P and $^{19}$F have I $= \frac{1}{2}$ and are 100% naturally abundant.

PF$_3$Cl$_2$ is trigonal bipyramidal and exists as 3 possible isomers:

| **A** | **B** | **C** |
|---|---|---|
| 2 chemically different F | 2 chemically different F | 1 type of F |

At -90 °C there are two $^{19}$F signals so structure **C** can be excluded. The $^{19}$F NMR spectra of **A** and **B** exhibit two $^{19}$F signals in the ratio 1:2.

In **B**:
F$_a$ doublet (due to P-F coupling) of triplets (due to coupling to 2 other F).
F$_x$ doublet (due to P-F coupling) of doublets (due to coupling to 1 other F).

In **A**: the reverse.

At ambient temperatures, Berry Pseudorotations interchange F$_a$ and F$_x$ rapidly on the NMR timescale giving rise to a single, time averaged signal which is a doublet due to P–F coupling and which is maintained because the fluxionality does not involve P–F cleavage.

We cannot distinguish between the structures from NMR. However, as Cl is larger than F, then from VSEPR rules, Cl will prefer the equatorial sites and hence structure **A** is correct.

S20

# Interpretation of the $^{13}C$ NMR spectrum of $[Rh_4(CO)_{12}]$ at different temperatures

**Problem**

Explain why the $^{13}C$ NMR spectrum of $[Rh_4(CO)_{12}]$ in solution consists of

■ a single broad peak at room temperature;
■ a sharp quintet at 60 °C; and
■ four signals (three doublets and one triplet) at -80 °C.

NOTE: Rh has spin number, $I = \frac{1}{2}$

**Prior knowledge**
Structures of rhodium complexes; nuclear magnetic resonance spectroscopy.

**This problem is suitable for:**
■ Second year students
■ Tutorial – group work
■ Approximately 45 minutes

**Knowledge/skills gained**
Improved ability to deduce and explain structures of compounds from $^{13}C$ NMR spectroscopic information.

## T20

# Interpretation of the $^{13}$C NMR spectrum of [Rh$_4$(CO)$_{12}$] at different temperatures

Ask students to draw the possible structures of [Rh$_4$(CO)$_{12}$] in its solid state and then eliminate those that do not correlate with the spectroscopic data.

**Solution**

In the solid state, [Rh$_4$(CO)$_{12}$] has the C$_{3v}$ structure shown below:

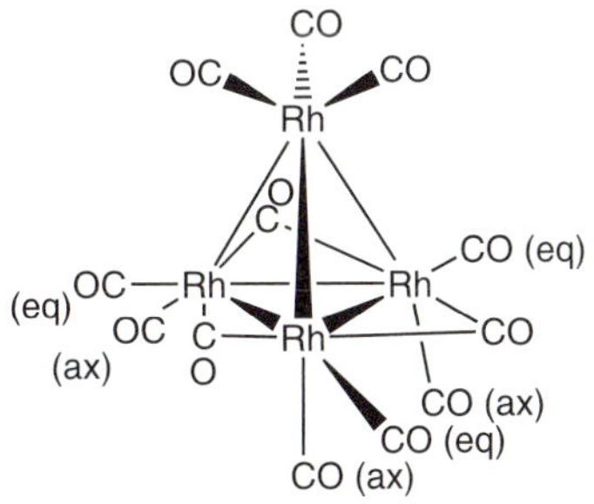

The $^{13}$C NMR spectrum at -80 °C is consistent with the structure shown above. The three equivalent carbonyl ligands of the Rh(CO)$_3$ group give a doublet due to coupling with one rhodium atom ($^{103}$Rh has spin $\frac{1}{2}$). The axial (ax) and equatorial (eq) carbonyl ligands of the three equivalent Rh(CO)$_2$ fragments give the remaining two doublets observed. The triplet may be assigned to the three equivalent bridging carbonyl ligands which couple with two $^{103}$Rh atoms.

At room temperature the observation of a single broad carbonyl resonance is consistent with the onset of a dynamic process which is beginning to make all the carbonyl ligands equivalent on the NMR scale. This could involve interconversion of terminal and bridging carbonyl ligands, and pairwise exchange of carbonyl ligands between adjacent rhodium atoms.

At 60 °C carbonyl exchange between metal centres has become rapid on the NMR scale and the quintet splitting reflects coupling with all four rhodium atoms.

# Identification of a drug in racehorses

**Problem**

As a trained analytical chemist you have been given the task of analysing a number of urine samples belonging to several racing horses. The samples all appear to be clear until you come to the final one, which seems to contain traces of a pain-killing drug.

You manage to isolate a crude sample of the drug and you suspect it to be 4-butyl-3,5-dioxo-1,2-diphenylpyrazolidine.

1.  What techniques could you use to purify the sample?

2.  How could you check that the sample was now pure?

3.  What could you do to confirm your suspicions?

4.  Using the information you have, predict the $^{13}$C NMR spectrum for the drug giving approximate chemical shift values and indicating any splitting due to coupling with hydrogen atoms.

5.  In the infrared spectrum will the presence of nitrogen move the C=O stretch absorbance to a higher or lower frequency than 1700 cm$^{-1}$?

**Prior knowledge**
Principles of sample analysis; $^{13}$C NMR and the effects of different species on chemical shifts.

**This problem is suitable for:**
■ Second year students
■ Tutorial – group work or class discussion
■ Approximately 40 minutes

**Knowledge/skills gained**
The application of known information in a practical sense. Confidence to suggest ideas in brainstorming sessions.
Students will develop an idea of a real life situation when all aspects of analytical chemistry are linked together to solve a problem. They will also gain practice predicting $^{13}$C NMR spectra for a given structure.

# Identification of a drug in racehorses

Depending on the level of difficulty required for the problem, the structure could be given. Question 1 could be extended to discuss each technique.

**Solution**

1. Techniques which could be used to purify the sample include:
   ■ Gas liquid chromatography;
   ■ High performance liquid chromatography; and
   ■ Crystallisation.

2. Techniques which could be used to test the purity of the sample include:
   Melting point – pure compounds have a sharp melting point. Impurities lower the melting point and the compound will melt over a range.
   Thin layer chromatography – pure substances should produce a single spot.
   Gas chromatography – used for volatile substances. A pure compound produces a single peak.
   Microanalysis.

3. The drug is suspected to be 4-butyl-3,5-dioxo-1,2-diphenylpyrazolidine. To confirm this obtain infrared, NMR and mass spectra of the sample and compare with authentic spectra of the drug.

4.

| Carbon | Approximate $\delta$/ppm | Splitting pattern |
|--------|--------------------------|-------------------|
| a | 20 | quartet |
| b | 30 | triplet |
| c | 30 | triplet |
| d | 30 | triplet |
| e | 45 | doublet |
| f | 170 | singlet |
| g | 110 | singlet |
| h | 130 | doublet |
| i | 130 | doublet |
| j | 130 | doublet |

5. The presence of the nitrogen atom causes the carbonyl frequency to shift to a lower frequency. The lone pair of electrons on nitrogen interacts with the carbonyl group. This results in less energy being required to stretch the bond and therefore a lower frequency.

## S22

# Identifying unknown compounds from spectral data

**Problem**

Identify the organic compounds, **A–F**, from the following data.

(i)  **A** has an infrared spectrum which indicates the presence of an O–H bond and it contains only carbon, hydrogen and oxygen. Its molar mass is 74 g mol$^{-1}$. The $^{13}$C NMR spectrum (fully decoupled) of **A** shows only two peaks (one large and one small). **A** is resistant to oxidation with potassium dichromate.

(ii)  **B** has a molar mass of 74 g mol$^{-1}$ and shows an O–H absorption in its infrared spectrum. On oxidation it yields an acidic compound **C** of molar mass 88 g mol$^{-1}$. Both **B** and **C** have three signals in their $^{13}$C NMR spectra.

(iii) **D** is an isomer of **A** and **B** but does not have an O–H stretch in its infrared spectrum. The $^{13}$C NMR spectrum contains four signals. Reaction of **D** with hydrobromic acid (aqueous HBr) yields a mixture of two bromides, **E** with a molar mass of 94/96 g mol$^{-1}$ and **F** with a molar mass of 122/124 g mol$^{-1}$.

**Prior knowledge**
$^{13}$C nuclear magnetic resonance spectroscopy; reactions of alcohols.

**This problem is suitable for:**
- Second year students
- A tutorial – group or individual work
- Approximately 30 minutes

**Knowledge/skills gained**
Deduction of structures/compounds from chemical and spectroscopic data.
Interpretation of $^{13}$C NMR spectra.

# Identifying unknown compounds from spectral data

A suggested approach for this problem is for the students to rewrite each part of the question, since a lot of information is given. By sorting out the relevant items the exercise will be much easier to follow and answer.

An example is shown below.

**Solution**

(i) The molar mass of **A** is 74 g mol$^{-1}$. **A** is most likely to be an alcohol due to the presence of an O–H stretch in the IR spectrum. Since **A** is resistant to oxidation it is probably a tertiary alcohol.
The $^{13}$C NMR spectrum has two peaks – one large and one small peak. The large peak is possibly due to the three alkyl groups of a tertiary alcohol. The small peak is due to the remaining carbon.
The simplest tertiary alcohol is 2-methylpropan-2-ol (*t*-butanol) that has a molar mass of 74 g mol$^{-1}$. Therefore **A** is

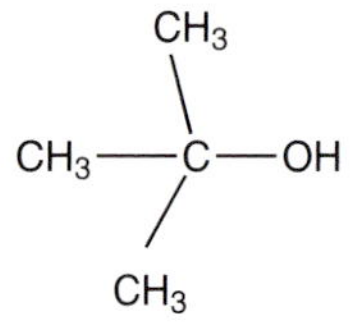

$$CH_3-\overset{\overset{\displaystyle CH_3}{|}}{\underset{\underset{\displaystyle CH_3}{|}}{C}}-OH$$

(ii) **B** is also an alcohol, confirmed by the O–H stretch in the IR spectrum. **B** is oxidised to **C**. **B** has three signals in the $^{13}$C NMR spectrum. Therefore, there are carbon atoms in three different chemical environments. **B** has the same molar mass as **A**, so it is likely to be an isomer of **A**. Possible isomers of **A** are $CH_3CH_2CH_2CH_2OH$ (1) and $(CH_3)_2CHCH_2OH$ (2).

(1) has four signals in its $^{13}$C NMR spectrum, therefore **B** is most likely to be (2) since the methyl groups are equivalent. Therefore **B** is

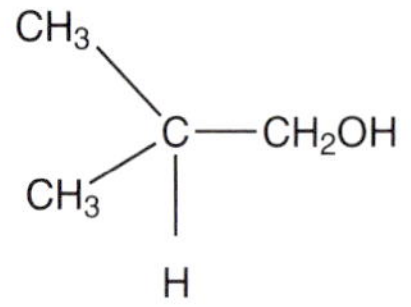

$$\overset{\displaystyle CH_3}{\underset{\displaystyle CH_3}{\diagdown}}\overset{|}{\underset{|}{C}}-CH_2OH$$
$$\underset{\displaystyle H}{}$$

C is the result of oxidising **B**. Therefore **C** is $(CH_3)_2CHCO_2H$,
which has a molar mass of 88 g mol$^{-1}$ and three signals in the $^{13}$C NMR
spectrum. Therefore **C** is

(iii) **D** is not an alcohol because it does not have an O–H stretch in the IR
spectrum. The four carbons are in chemically different environments since
there are four signals in the $^{13}$C NMR spectrum. **D** is an isomer of **A**. **D** is
most likely an ether and is $CH_3OCH_2CH_2CH_3$ (3) or $CH_3OCH(CH_3)_2$ (4).

The $^{13}$C NMR spectrum of (4) does not have four signals. Therefore **D** is (3)
*ie* 1-methoxypropane.

Bromination of **D** yields bromides *ie*

$$CH_3OCH_2CH_2CH_3 \xrightarrow{HBr} CH_3Br + CH_3CH_2CH_2Br + CH_3OH + CH_3CH_2CH_2OH$$

$$\phantom{CH_3OCH_2CH_2CH_3 \xrightarrow{HBr} } \textbf{E} \qquad\qquad \textbf{F}$$

## S23

# Compound identification from chemical reactions and NMR spectra

**Problem**

### Part A

Deduce the structures of compounds **A–D** given the following information. Briefly discuss the mechanisms of the reactions **A → B** and **C → D**.

(i)   Vigorous oxidation of a hydrocarbon **A**, $C_7H_{12}$, gave heptane-2,6-dione.

(ii)  Treatment of **A** with osmium tetroxide gave a compound **B**, $C_7H_{14}O_2$.

(iii) **A** reacted with perbenzoic acid to give a product **C**, $C_7H_{12}O$.

(iv) **C** on hydrolysis with dilute aqueous sulfuric acid formed **D**, $C_7H_{14}O_2$.

### Part B

Explain how the three isomers of dimethylbenzene could be distinguished using $^{13}C$ NMR spectroscopy.

Explain how the three isomers of trimethylbenzene could be identified from the methyl signals in their $^1H$ NMR spectra?

### Prior knowledge

Isomerism and spectroscopy.

### This problem is suitable for:

■   Second year students
■   A tutorial – groups of four students
■   Approximately 1 hour

### Knowledge/skills gained

Improved skills in identifying relevant information from a collection of data. Application of $^1H$ and $^{13}C$ NMR spectroscopy.

# Compound identification from chemical reactions and NMR spectra

A suggested approach for part A is to divide students into groups of four. Each student is given a part of the problem to solve individually. Each group should then discuss their proposed structures.

**Solution**

**Part A (i)–(iii)**

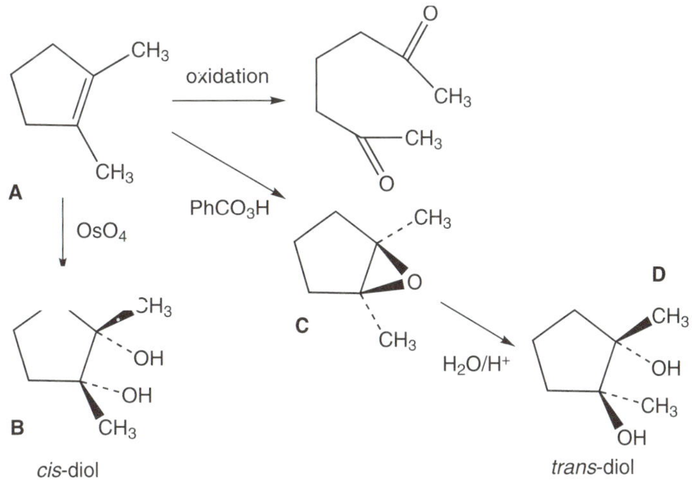

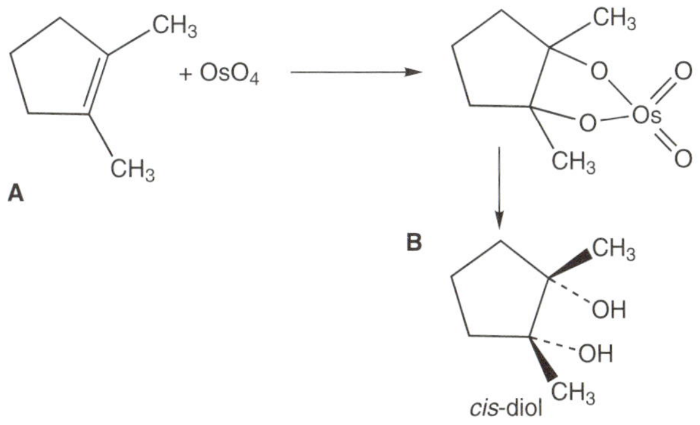

(iv)

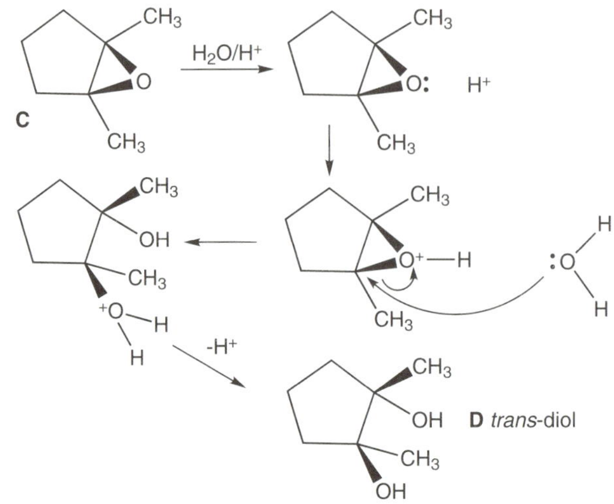

**Part B**

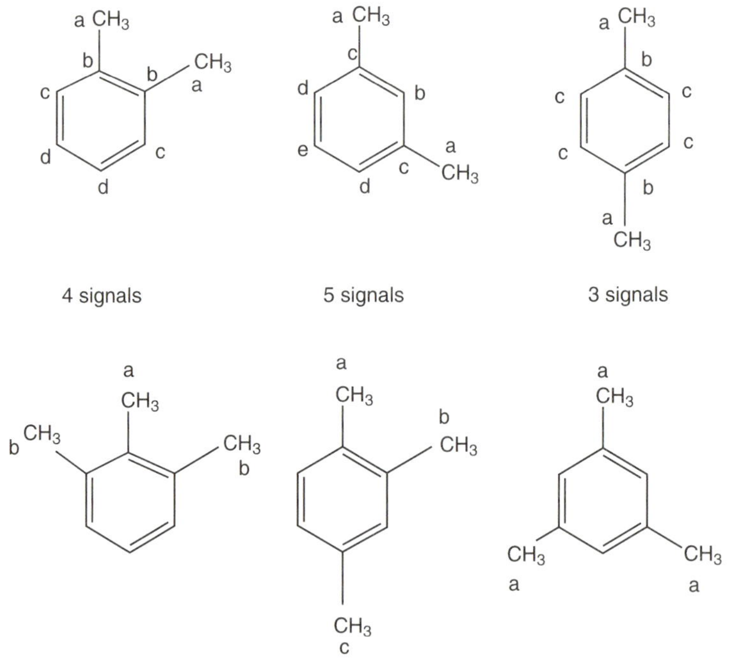

4 signals          5 signals          3 signals

2 signals          3 signals          1 signal

$^{13}$C NMR spectra usually show a series of peaks, each representing a carbon atom in a different environment. The methyl groups exhibit different signals in the $^1$H NMR spectrum depending on their neighbouring atoms.

# S24

# Identifying unknown organic compounds (I)

**Problem**

### Part 1

Compound **A**, $C_{20}H_{28}O_2$, reacts with sodium hydroxide followed by acidification to give an acidic compound **B**, $C_{20}H_{30}O_3$. On mild hydrogenation, **B** gives $C_{20}H_{32}O_3$, but **B** does not give a crystalline derivative with 2,4-dinitrophenylhydrazine.
Deduce the structure of **B**.

Return to your groups and discuss your proposed structure(s) for the compound(s) you have been designated to solve. Outline your reasoning to the other group members.
As a group propose structures for all the compounds **A–E**.

### Part 2

Ozonolysis of **B**, $C_{20}H_{30}O_3$, gives **C**, $C_{12}H_{22}O_3$, and compound **D**, $C_8H_8O_2$.
**D** forms a crystalline derivative with 2,4-dinitrophenylhydrazine and a violet colour with neutral, aqueous iron(III) chloride.
Deduce the structures of compounds **B** and **C**.

Return to your groups and discuss your proposed structure(s) for the compound(s) you have been designated to solve. Outline your reasoning to the other group members.
As a group propose structures for all the compounds **A–E**.

### Part 3

Mild oxidation of **C** gives a diacid **E**, $C_{12}H_{22}O_4$, which shows six signals in the proton decoupled $^{13}C$ NMR spectrum; the off-resonance decoupled spectrum shows five triplets and a deshielded singlet.
Deduce the structure of compound **E**.

Return to your groups and discuss your proposed structure(s) for the compound(s) you have been designated to solve. Outline your reasoning to the other group members.
As a group propose structures for all the compounds **A–E**.

### Part 4

The proton NMR spectrum of **D**, $C_8H_8O_2$, has the following signals:
$\delta7.8$ (2H, doublet, J = 8.00 Hz), $\delta7.5$ (1H, broad singlet, undergoes $D_2O$ exchange), $\delta6.9$ (2H, doublet, J = 8.00 Hz) and $\delta2.5$ (3H, singlet).
Deduce the structure of compound **D**.

Return to your groups and discuss your proposed structure(s) for the compound(s) you have been designated to solve. Outline your reasoning to the other group members.
As a group propose structures for all the compounds **A–E**.

**Prior knowledge**
Organic chemical reactions; $^{1}$H and $^{13}$C nuclear magnetic resonance spectroscopy.

**This problem is suitable for:**
■ Second year students
■ A tutorial – individual work and then group work (preferably four students per group)
■ Approximately 1 hour

**Knowledge/skills gained**
Deduction of compounds/structures from chemical and spectroscopic information.
Team working.

# T24

# Identifying unknown organic compounds (I)

This problem is self-explanatory. Students should be split up into small groups, ideally of four. Each student in the group should be given a different part of the problem. Students should propose structures for the part they have been given and discuss their reasoning in groups.

The students will probably have difficulty deducing structures for their individual parts, but even by writing down some ideas they are showing that they are thinking, *eg* the test for an aldehyde and ketone. Then when they pool their ideas, brainstorming will lead to the solutions.

**Solution**

Compound **A**, $C_{20}H_{28}O_2$, when compared with $C_{20}H_{42}$, is found to have 7 Double Bond Equivalents (DBE).

**B** is formed when water is added to **A**. We know **B** is not an aldehyde or a ketone as it has no reaction with 2,4-dinitrophenylhydrazine.
The structure of **B** could be 'deduced' here.

The C=C bond in **B** is confirmed by ozonolysis when it is replaced by 2 C=O. (This will be discovered on discussion within the group).

**D** is an aldehyde or ketone and is also a phenol (violet coloured complex with $Fe^{3+}$).

**C** is oxidised easily with addition of one oxygen to give **E**.
As **E** contains only six signals in the $^{13}C$ NMR, it must be a symmetrical molecule and the deshielded singlet must be due to $CO_2H$ groups; the five triplets are due to $CH_2$ groups.

Hence, the structure of **E** is $HO_2C(CH_2)_{10}CO_2H$
and therefore **C** is $HO_2C(CH_2)_{10}CHO$

The proton NMR spectrum of **D**, $C_8H_8O_2$ (5 DBE) is consistent with an OH group ($\delta7.5$), confirming a phenol; there must be a C=O group and the absence of a signal at $\delta9–10$ (aldehyde) shows that it must be a ketone. This correlates with the singlet integrating for 3H, so it is probably a methyl ketone $COCH_3$.

**D** probably contains an aromatic ring and the two doublets suggest 1,4- substitution.

Hence the structure of **D** is:

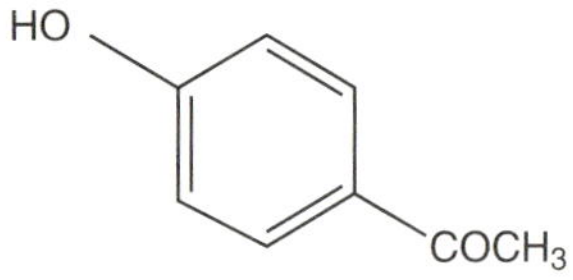

Working backwards the structure of **B** can be deduced – there is a C=C where the aldehyde and ketone are present in **C** and **D**.

Hence, **B** is:

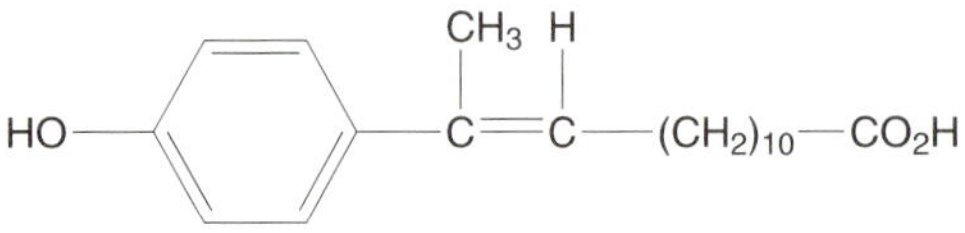

and as **A** is reformed by dehydration of **B**, **A** must be:

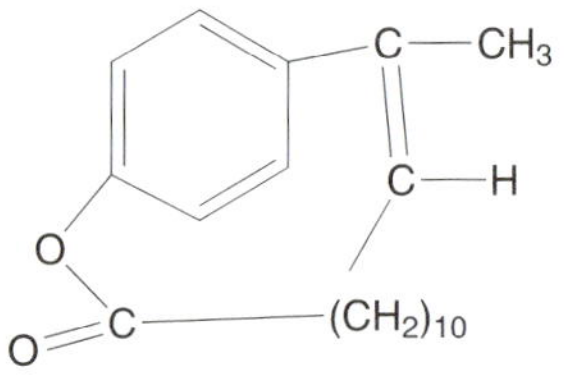

**A** is a cyclic ester (lactone) which readily undergoes hydrolysis.

# S25

# Identifying unknown organic compounds (II)

**Problem**

Compound **A**, $C_{13}H_{14}$, forms a tetrabromo derivative, $C_{13}H_{14}Br_4$, when treated with bromine in tetrachloromethane. On reaction with ozone followed by oxidative work-up **A** gives two products, benzoic acid and compound **B**, $C_6H_8O_5$.

**B** gives a red precipitate with 2,4-dinitrophenylhydrazine and on gentle warming **B** is converted to compound **C**, $C_5H_8O_3$.

Although **B** fails to give iodoform with iodine in the presence of sodium hydroxide, **C** reacts to yield iodoform and compound **D**, $C_4H_6O_4$, after acidification. On heating to its melting point **D** is converted into compound **E**, $C_4H_4O_3$.

The off-resonance decoupled $^{13}C$ NMR spectrum of **E** displays only two resonances: a singlet at 178 ppm and a triplet at 29 ppm. The off-resonance decoupled $^{13}C$ NMR spectrum of **B** displays six resonances: three singlets in the 175–210 ppm carbonyl region and three triplets in the 30–60 ppm aliphatic region.

From the above information deduce structures for compounds **A–E**.

**Prior knowledge**
Organic reactions; nuclear magnetic resonance spectroscopy and structure elucidation.

**This problem is suitable for:**
■ Second year students
■ A tutorial – group work (2–3 students) followed by class discussion
■ Approximately 1 hour

**Knowledge/skills gained**
Reactions of alkenes and ketones.

# Identifying unknown organic compounds (II)

This problem involves identifying structures from information provided. A suggested approach has been given, demonstrating the use of diagrams in effective problem solving. The diagram shows students the steps to each compound.

**Solution**

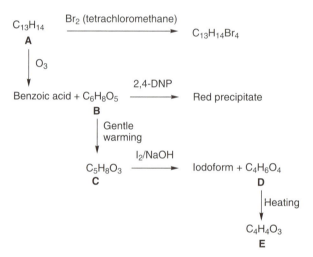

**A** contains seven Double Bond Equivalents, DBE, and two C=C double bonds (tetrabromo derivative).

Ozonolysis should cleave both C=C double bonds. As only two products are formed, one of the double bonds is probably in a ring. Thus, we account for three of the seven DBE.
(Oxidative work-up in ozonolysis converts any aldehyde groups to carboxylic acid groups).

**B** is $C_6H_8O_5$ so it has three DBE, probably three C=O groups (at least one of which is a ketone; reacts with 2,4-dinitrophenylhydrazine and unlikely to be an aldehyde at this stage).

**C** is formed by loss of carbon dioxide from **B** – such an easy loss is probably due to a β-ketoacid.
**C** contains a $CH_3CH(OH)$ or a $CH_3CO$ unit (iodoform test) and **B** does not.

**D** is $C_4H_6O_4$ (two DBE) – probably a dicarboxylic acid, which is dehydrated on heating.
(Students might 'guess' the structure of **E** at this stage).

The $^{13}C$ NMR spectrum of **E** shows that it is a symmetrical molecule containing deshielded quaternary carbons and deshielded CH groups.

Hence,

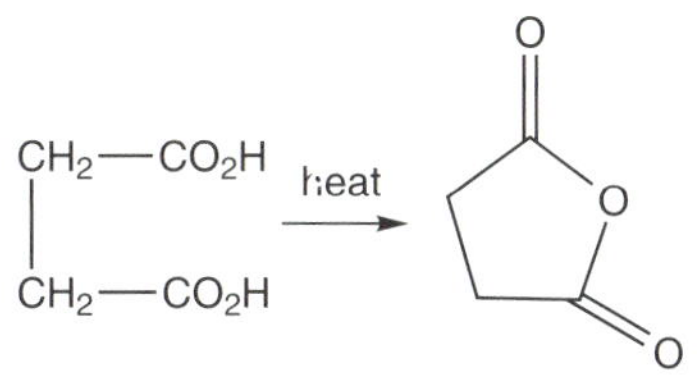

D          E

**B** contains three deshielded quaternary carbons (ketone or carboxylic acid) and three $CH_2$ groups. The reaction sequence is

$$-COCH_2CO_2H \text{ (in } \mathbf{B}) \rightarrow -COCH_3 \text{ (in } \mathbf{C}) \rightarrow -CO_2H \text{ (in } \mathbf{D}) + CHI_3$$

Hence **C** is $CH_3COCH_2CH_2CO_2H$ and **B** is $HO_2CCH_2COCH_2CH_2CO_2H$.

Possible structures for **A** have a PhCH= instead of a C=O group in **B** with the remaining two C=O groups forming a ring with a C=C bond. Attaching the PhCH group in turn from the left gives three possibilities for **A**.

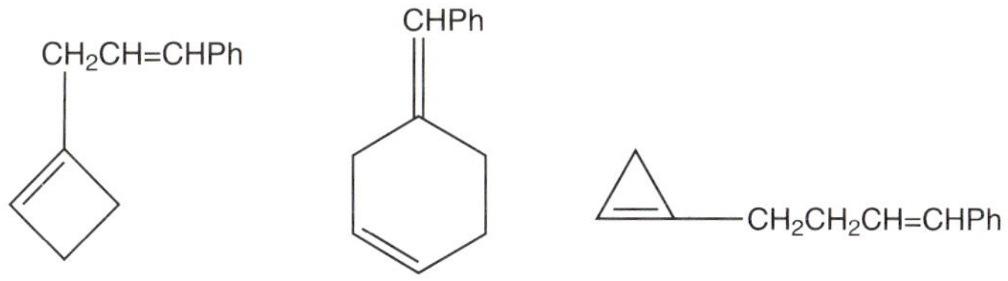

**Acknowledgement**
Dr T W Bentley and Dr J A Ballantine, University of Wales, Swansea

# Identifying unknown organic compounds (III)

**Problem**

An optically active compound **A**, $C_9H_{10}O_3$, when synthesised in its racemic form can be resolved using a chiral amine.

Three signals were observed in the $^1H$ NMR spectrum ($D_2O$ solution) of **A**: $\delta 7.30$ (5H, singlet), $\delta 4.00$ (1H, triplet), and $\delta 3.60$ (2H, doublet).

Compound **B**, $C_{10}H_{12}O_3$, is a neutral derivative of **A**. The $^1H$ NMR spectrum ($D_2O$ solution) of **B** exhibited an additional signal at $\delta 3.80$ (3H, singlet).

**B** reacts readily with a chiral isocyanate **C**, (R)-$C_8H_9N{=}C{=}O$, to form a urethane derivative **D**, $C_{19}H_{21}NO_4$. Mild oxidation of **B** produces an aldehyde. The $^1H$ NMR spectrum ($D_2O$ solution) of **D** had all the signals showing its relationship with **B** and signals at $\delta 7.15$ (5H, singlet), $\delta 3.00$ (1H, quartet), $\delta 1.50$ (3H, doublet).

Suggest, giving your reasoning, structures for **A**, **B** and **D**.

**Prior knowledge**
$^1H$ NMR spectroscopy.

**This problem is suitable for:**
- Second year students
- A tutorial – group work
- Approximately 45 minutes

**Knowledge/skills gained**
Interpretation of $^1H$ NMR spectra and use in structure elucidation.
Use of $D_2O$ in $^1H$ NMR spectroscopy.
Resolution of racemic compounds.

# T26

# Identifying unknown organic compounds (III)

An initial discussion of racemic compounds and methods of resolution might be useful to students.

Students may need some guidance in interpreting the NMR spectra.

**Solution**

*Compound A*

From the information given, **A** is probably a carboxylic acid since it can be resolved using a chiral amine.

The $^1$H NMR spectrum was run in $D_2O$, therefore signals for exchangeable protons are not seen. The number of double bond equivalents for **A** is 5. The following deductions can be made from the $^1$H NMR spectrum:

■ A singlet at δ7.30 integrating for 5H suggests the presence of an aromatic ring – probably a mono-substituted benzene ring.

■ A triplet at δ4.00 which integrates for 1H indicates the presence of a $CH$–$CH_2$ group. This signal has been shifted downfield possibly due to the presence of a carboxylic acid group.

■ A doublet at δ3.60 indicates the presence of a $CH_2$ group. This signal integrates for 2H therefore suggesting $CH$–$CH_2$. This signal has also been shifted downfield due to the presence of a carboxylic acid group.

From the $^1$H NMR spectrum the structure can be built up:

$PhCHXCH_2Y$

This accounts for $C_8H_8$. The molecular formula is $C_9H_{10}O_3$. There are two hydrogens unaccounted for in the $^1$H NMR spectrum – both of these must be exchangeable protons – *ie* OH. From the information given, **A** is known to be a carboxylic acid *ie*

$PhCHXCH_2Y$ where X or Y = $CO_2H$

This accounts for $C_9H_9O_2$ leaving OH. From NMR chemical shift tables the structure of **A** is therefore:

$PhCHCO_2H$
|
$CH_2OH$

## Compound B

The $^1$H NMR spectrum of **B** is the same as **A** with an additional peak at
$\delta 3.80$ which integrates for 3H. From NMR chemical shift tables this peak is
probably due to a methoxyl group. **B** must therefore be the methoxy ester
derivative of **A**:

PhCH(CO$_2$CH$_3$)CH$_2$OH

## Compound D

**D** is formed by reaction of **B** and **C** and is a urethane. The following
deductions can be made from the $^1$H NMR spectrum:

- $\delta 7.30$ (5H, s)    mono-substituted benzene ring (from **B**)

CH$_3$            Ph
|                |
PhCHNHCOOCH$_2$CHCO$_2$CH$_3$

- $\delta 7.15$ (5H, s)    mono-substituted benzene ring
- $\delta 4.00$ (1H, t)    C*H*CH$_2$ (from **B**)
- $\delta 3.80$ (3H, s)    OC*H*$_3$ (from **B**)
- $\delta 3.60$ (2H, d)    C*H*$_2$CH (from **B**)
- $\delta 3.00$ (1H, q)    C*H*CH$_3$ (from splitting pattern)
- $\delta 1.50$ (3H, d)    CHC*H*$_3$ (from splitting pattern)

**D** is therefore:

### Acknowledgement
Dr T W Bentley and Dr J A Ballantine, University of Wales, Swansea

## S27

# Elephant poaching

**Problem**

Elephant poaching is rife in Africa. There is global concern over the ever-decreasing population of the African elephant. This has influenced many countries in Africa to disallow the export of ivory. Some countries, however, still allow it. For this reason it is necessary to establish where the ivory originated.

As a chemist you have been given the job of devising an analytical procedure that will determine an elephant's place of origin by analysis of its ivory tusks.

Your answer should include the reasoning behind your choice of analytical technique.

### Prior knowledge
A degree of knowledge is necessary on analytical techniques (mass spectrometry in particular if the suggested solution is used). In this case an understanding of isotopes is also essential.

### This problem is suitable for:
■ First and second year students
■ A tutorial – group work
■ Approximately 1 hour

### Knowledge/skills gained
An improved ability to apply techniques learnt in theory to actual events. Searching literature for relevant information and group working.

T27

# Elephant poaching

This problem is probably best approached by groups. Discussion between members should be encouraged and a report could be submitted for assessment. Some research may be required, especially when considering the differences between the isotopes.

A suggested solution is outlined below.

**Solution**

The ivory tusks can be analysed to determine the diet of the elephant and from this, its dwelling place.

Mass spectrometry could be used to determine the amounts of carbon, nitrogen and strontium in the ivory tusks.

The different ratios of $^{13}C$ to $^{12}C$ should indicate whether the elephant is from a wood or grassland area. This is due to the differences in glucose production. Trees have smaller $^{13}C$ to $^{12}C$ ratios than grass. Each time a carbon atom is added to build up a simpler compound to give a more complex product $^{12}C$ is preferred as it reacts faster. Because trees require more steps to build up glucose they have a smaller $^{13}C$ to $^{12}C$ ratio. This is shown in the elephant's tusks. The ratios between $^{15}N$ and $^{14}N$ and $^{87}Sr$ and $^{86}Sr$ could also be used.

# S28

# Using spectroscopic data

**Problem**

### Part 1

When compound **A** is treated with $N_2O_5$ in dichloromethane, two isomeric compounds **B** and **C** are formed. Using the spectra below propose a structure for compound **A** and explain how this is consistent with the information provided.

Composition: C 90.5%; H 9.5%

Mass spectrum: major peaks at m/z 106 ($M^+$), 105, 91, 77, 65, 51

$^1H$ NMR spectrum (recorded at 90 MHz):

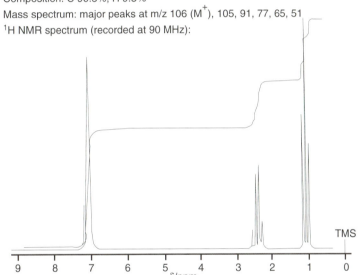

$^{13}C$ NMR spectrum (fully decoupled). Letters in parentheses refer to the multiplicity of the appropriate peak in the off-resonance spectrum (s - singlet, d - doublet, t - triplet, q - quartet).

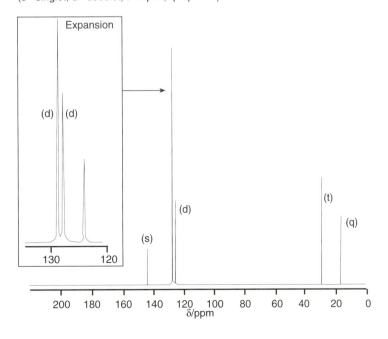

**Part 2**

When compound **A** is treated with $N_2O_5$ in dichloromethane, two isomeric compounds **B** and **C** are formed.

Analyse the mass spectral and NMR information for **B**, and on this basis deduce its structure. Suggest a mechanism for the formation of this compound.

Mass spectrum: major peaks at m/z 151 ($M^+$), 136, 121, 105, 79, 77, 51

$^1H$ NMR spectrum (recorded at 90 MHz):

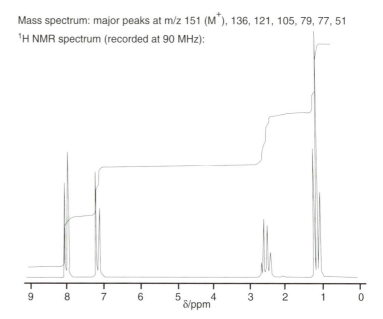

$^{13}C$ NMR spectrum (fully decoupled):

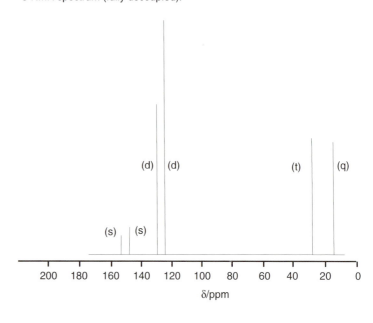

**Part 3**

When compound **A** is treated with $N_2O_5$ in dichloromethane, two isomeric compounds **B** and **C** are formed in good yield.

What structure would you expect for isomer **C**?

How does the structure you propose for **C** account for:

(i)  The appearance of the proton NMR spectrum shown?

(ii) The observation that the base peak in its mass spectrum has m/z 134?

$^1$H NMR spectrum (recorded at 90 MHz):

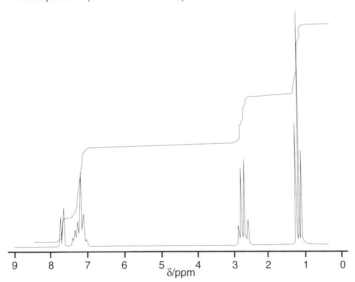

Portion of the $^1$H NMR spectrum (recorded at 300 MHz):

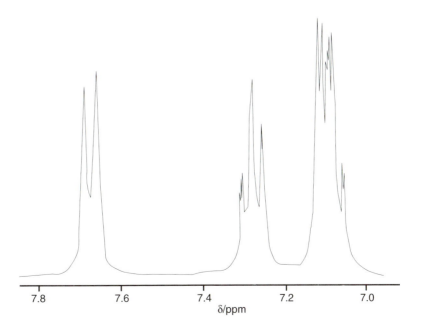

**Prior knowledge**

Spectroscopic analysis; mass spectra, $^{1}$H NMR and $^{13}$C NMR; aromatic chemistry.

**This problem is suitable for:**

■ Second year students
■ A tutorial – group work
■ Approximately 1 hour

**Knowledge/skills gained**

Electrophilic aromatic substitution.
Improved spectroscopy skills.
Group work.

# T28

# Using spectroscopic data

This problem has been split into three parts. There are various ways in which the problem could be approached. One way is by splitting the class into groups of 3 or 6 and giving each member (or in the case of 6 each partner) a part of the problem to try and solve in a given time. Then they can rejoin as a group to discuss their theories and solve the problem fully. Another way could be to solve the problem as a whole, either in a class discussion or as group work.

**Solution**

**Part 1**
**Compound A**
Composition:
Carbon       = composition/molar mass
               = $90.50/12.00$ g mol$^{-1}$ = 7.50

Hydrogen   = composition/molar mass
               = $9.50/1.00$ g mol$^{-1}$ = 9.50

Divide by smallest number:
Carbon      = 7.50/7.50    = 1.00
Hydrogen   = 9.50/7.50    = 1.27
Multiply by four to give whole numbers. This gives a ratio of approximately four carbon atoms to five hydrogen atoms.
Therefore, we should expect $C_4H_5$ as the empirical formula.

**$^1$H NMR**
The spectrum shows a multiplet at 7.00 ppm. This corresponds to aromatic ring protons. A quartet is shown at 2.50 ppm, this will be for a $CH_2$ group adjacent to a $CH_3$ and the triplet at 1.00 ppm will be for a $CH_3$ group adjacent to a $CH_2$ group.

It is not clear at this stage how many of these aliphatic groups are present. However, integration gives a ratio of 5:2:3 therefore one $C_6H_5$, one $CH_2$ and one $CH_3$ are present and the actual molecular formula is $C_8H_{10}$. It can therefore be presumed that the structure of **A** is:

CH₂CH₃

**$^{13}$C NMR (fully decoupled)**
As in the proton NMR spectrum, the signals upfield indicate the presence of aliphatic groups. The peak at 15.00 ppm (quartet) indicates a $CH_3$ group and the line at 30.00 ppm (triplet) indicates a $CH_2$ group.
The three doublets plus one singlet indicate a monosubstituted benzene ring.

Using the proposed structure from the analysis of $^1$H NMR:

Each carbon in the aromatic ring, in a different chemical environment, is given a different label (eg a–d).

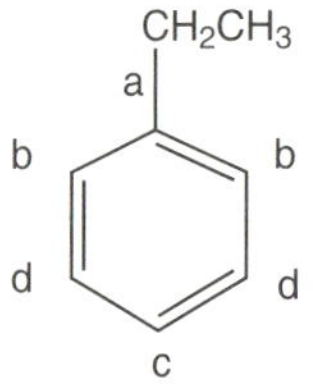

CH$_2$CH$_3$

a

b          b

d          d

c

This corresponds to the spectrum, that is:

Carbon a = singlet
Carbon b = doublet
Carbon c = doublet
Carbon d = doublet

**Mass Spectrum**
Using the information from the mass spectrum we can check the molecular formula.

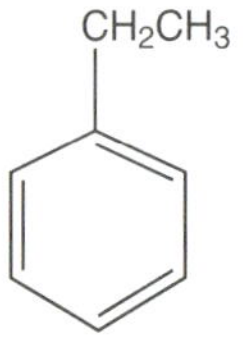

CH$_2$CH$_3$

m/z = 106

The remaining peaks correlate with the fragmentation of the proposed compound,

eg   91 = $C_6H_5CH^{2+}$
      77 = $C_6H_5^+$

CH$_2$CH$_3$

**A =**

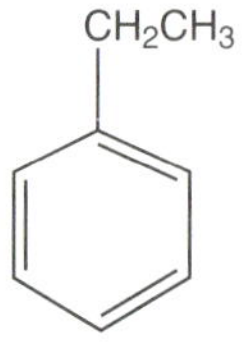

**Parts 2 and 3**
**Compounds B and C**

$N_2O_5$ (dichloromethane)

Compound **A** $\longrightarrow$ Compound **B** + Compound **C**

This reaction involves the nitronium ion. This can be shown by the information from the mass spectrum:

CH$_2$CH$_3$

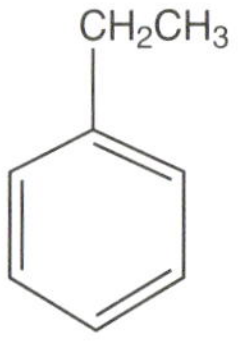

+   $NO_2^+$   (-H$^+$)  =  m/z 151

The mechanism for the reaction is:

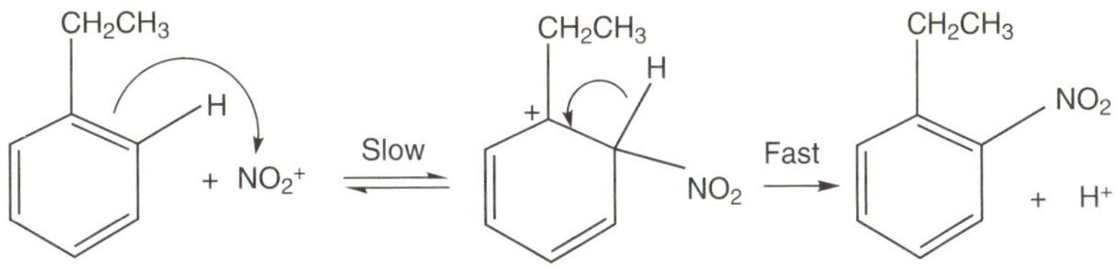

There could be 3 different products:

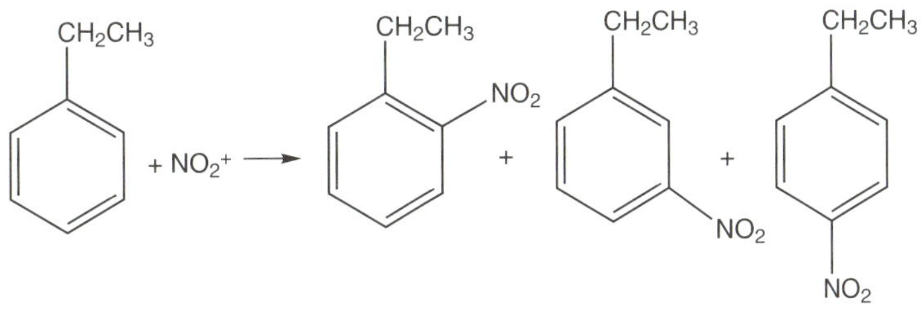

Ortho          Meta          Para

The products formed depend on whether the functional groups already present on the ring are electron-donating or electron-withdrawing. The ethyl group is electron-donating. Therefore, the ortho and para isomers are mainly formed. These deductions can be verified using the spectroscopic information provided.

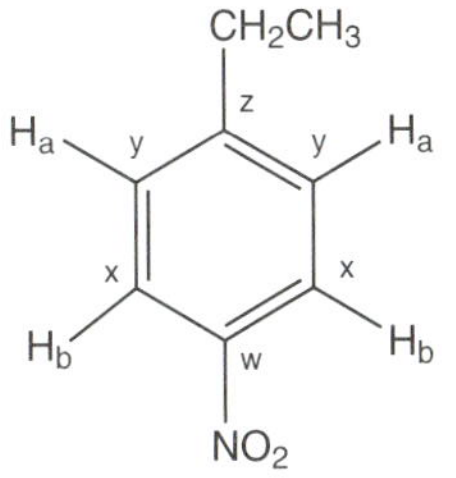

[1]H NMR for the aromatic ring Hs
There should be 2 signals:
Both doublets;
Both same intensity.

[13]C NMR for the aromatic ring Cs
There should be 4 signals:
2 singlets and 2 doublets;
z, w singlets and x, y doublets.

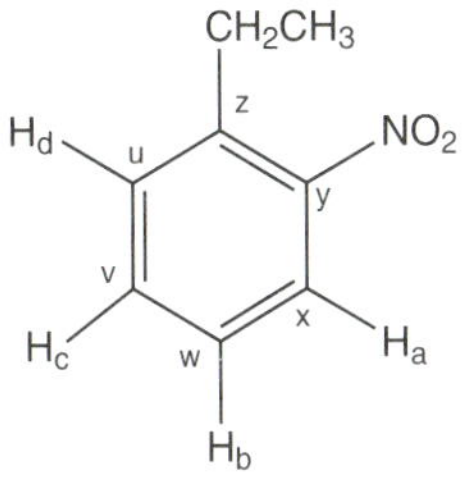

[1]H NMR for the aromatic ring Hs
There should be 4 signals:
All multiplets;
All same intensity.

[13]C NMR for the aromatic ring Cs
There should be 6 signals:
2 singlets and 4 doublets;
z, y singlets and x,w,v,u doublets.

Therefore, we can deduce that compounds **B** and **C** are:

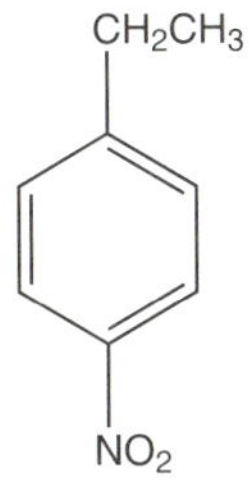

**B**

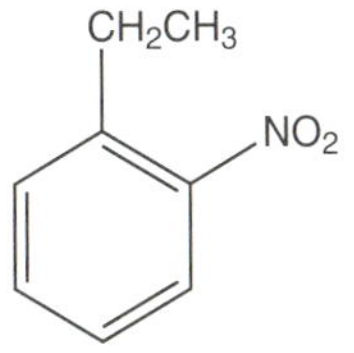

**C**

**Acknowledgement**
Wentworth College, University of York

# Predicting infrared and Raman active molecules

**Problem**

1. Of the following list of molecules, which will show infrared absorption spectra?

    (a)  $H_2O$      (b)  $CH_3Cl$      (c)  $N_2$      (d)  $H_2$
    (e)  $CH_3CH_3$    (f)  $CO_2$      (g)  HCl     (h)  $CH_4$

2. Of the following list of molecules, which will exhibit a rotational Raman spectrum?

    (a)  $CH_4$      (b)  $CH_3CH_3$    (c)  HCl
    (d)  $H_2$       (e)  $CH_3Cl$     (f)  $CH_2Cl_2$

**Prior knowledge**
Basic knowledge of the concepts behind both Raman and infrared
spectroscopy.

**This problem is suitable for:**
■ Second year students
■ A tutorial – group work
■ Approximately 30 minutes

**Knowledge/skills gained**
An understanding of the basic theory and applications of Raman and infrared
spectroscopy.

# T29

# Predicting infrared and Raman active molecules

This problem requires an understanding of the basic principles behind infrared and Raman spectroscopy. The questions are not too difficult and probably if lectures have not yet been covered on this subject a small amount of research by the students should enable the problem to be completed successfully and competently.

**Solution**

Raman spectroscopy involves changes in wavelength, and infrared spectroscopy involves emission or absorption of radiation. These techniques often show different spectra for the same molecules. The reason for this is that a molecule exhibiting bonds that are symmetrical (no dipole), for example $H_2$, will not be active in infrared, but the chances are that this molecule will be active in Raman, since it shows a change in polarisability.

1.  The following have a change in dipole and therefore give infrared absorption spectra:

    (a)        (b)        (e)        (f)        (g)        (h)

2.  The following have a change in polarisability and therefore give lines in Raman spectra:

    (b)        (c)        (d)        (e)        (f)

## S30

# Absorbance and transmittance of thiamine (Vitamin B1)

**Problem**

1.  During an assay of the thiamine (vitamin B1) content of a pharmaceutical preparation, the percent transmittance scale of the spectrophotometer was accidentally read instead of the absorbance scale.

    One sample gave a reading of 82.20% transmittance, and a second gave a reading of 50.70% transmittance, both at the wavelength of maximum absorbance.

    What is the ratio of concentrations of thiamine in the two samples?

2.  The absorbance of a $2.31 \times 10^{-5}$ mol dm$^{-3}$ solution of a compound is 0.82 at a wavelength of 266 nm in a 1.00 cm cell.

    Calculate the molar absorptivity (*ie* the molar decadic extinction coefficient) at 266 nm.

**Prior knowledge**
Formula for the Beer-Lambert Law and the relationship between absorbance and transmittance.

**This problem is suitable for:**
■   First year students
■   A tutorial – group work
■   Approximately 30 minutes

**Knowledge/skills gained**
Improved knowledge of absorbance calculations.

# Absorbance and transmittance of thiamine (Vitamin B1)

These problems should not require too much brainstorming between class members, although communication should be encouraged to stimulate ideas for tackling the problems successfully.

**Solution**

1.  As absorbance = $\log I_0/I$

    and transmittance = $I/I_0$ (x 100%) then:

    | | |
    |---|---|
    | 82.20% transmittance | = 0.82 |
    | $I_0/I$ | = 1.22 |
    | absorbance | = 0.09 |

    | | |
    |---|---|
    | 50.70% transmittance | = 0.51 |
    | $I_0/I$ | = 1.96 |
    | absorbance | = 0.29 |

    | | |
    |---|---|
    | Ratio | = 0.09/0.29 |
    | | = 0.31 |
    | | (1:3.23) |

2.  According to the Beer-Lambert Law, absorbance = $\varepsilon cl$

    where,
    $c$ = concentration
    $l$ = path length of radiation
    $\varepsilon$ = molar absorptivity

    therefore,

    $$0.82 = \varepsilon \times 2.31 \times 10^{-5} \text{ mol dm}^{-3} \times 1.00 \text{ cm}$$
    $$\varepsilon = 3.55 \times 10^4 \text{ dm}^3 \text{ mol}^{-1} \text{ cm}^{-1}$$

**Acknowledgement**
Dr John Smith, University of Newcastle upon Tyne

# Identification of compounds by thin layer chromatography

**Problem**

A laboratory assistant has been given three compounds. He is told that they are suspected to be the aromatic hydrocarbons known as anthracene, pyrene and phenanthrene. The assistant has also been supplied with the corresponding standards in order to run a thin layer chromatographic analysis.

The analysis shows that the three compounds match the three standards. This is shown by the $R_f$ (retardation factor) values on the chromatograms.

1. From the results obtained above would you conclude that the compounds are anthracene, pyrene and phenanthrene? Give reasons for your answer.

2. What other methods could be used to check the compounds' identities?

**Prior knowledge**
Chromatography techniques – gas, liquid, thin layer.

**This problem is suitable for:**
■ First year students
■ A tutorial – group work and discussion
■ Approximately 30 minutes

**Knowledge/skills gained**
Communication skills and aspects of chromatography.

# Identification of compounds by thin layer chromatography

**Solution**

1. This result should not be concluded as being correct.

   There are a lot of compounds similar to those suggested and it is possible they may have the same $R_f$ values. Therefore, this analysis alone does not confirm the compounds' identities.

2. The separate compounds could be analysed using spectroscopy – *eg* NMR, GCMS or infrared. Alternatively more analysis could be performed on other substances and the $R_f$ values compared. However, absolute identification cannot be achieved using comparison alone. Spectroscopy techniques must also be used.

# Spectrophotometry – application of the Beer-Lambert Law

**Problem**

A sample has an absorbance value of 0.45 at 280 nm and 0.59 at 260 nm. Assuming the path length to be 1 cm, calculate the concentration of each constituent of the sample.

The sample is made up of two components A and B and:

|   | 280 nm | 260 nm |
|---|--------|--------|
| A | $\varepsilon = 3.42 \times 10^{-7}\ dm^3\ cm^{-1}\ mol^{-1}$ | $\varepsilon = 1.06 \times 10^{-7}\ dm^3\ cm^{-1}\ mol^{-1}$ |
| B | $\varepsilon = 5.67 \times 10^{-6}\ dm^3\ cm^{-1}\ mol^{-1}$ | $\varepsilon = 1.70 \times 10^{-7}\ dm^3\ cm^{-1}\ mol^{-1}$ |

**Prior knowledge**
Knowledge of the Beer-Lambert Law is needed and an ability to solve simultaneous equations.

**This problem is suitable for:**
- Second year students
- A tutorial – group work
- Approximately 30 minutes

**Knowledge/skills gained**
Improved skills in using the Beer-Lambert Law.

## T32

# Spectrophotometry – application of the Beer-Lambert Law

This problem involves the testing of a student's mathematical and chemical abilities. Knowledge of the Beer-Lambert Law and recognition of the information given to form simultaneous equations, then correctly inserting these results into the equation to find concentration values.

**Solution**

As the Beer-Lambert Law is

| A | $= \varepsilon cl$ |
|---|---|
| Total Absorbance | = Absorbance of A + Absorbance of B |
| then, at 280 nm | $0.45 = 34.20\ A + 5.67\ B$ |
| and at 260 nm | $0.59 = 10.60\ A + 17.00\ B$ |

[Note: The above equations are all to $10^{-6}$.]
These equations can be solved simultaneously to give:

$$A = 8.30 \times 10^{-3} \qquad\qquad B = 2.90 \times 10^{-2}$$

Therefore, the concentrations are $\quad A = 8.30 \times 10^{-3}\ \text{mol dm}^{-3}$
and $\qquad\qquad\qquad\qquad\qquad B = 29.00 \times 10^{-3}\ \text{mol dm}^{-3}$

## S33

# Identifying rhodium carbonyl compounds from chemical and spectroscopic data

**Problem**

Suggest structures for the compounds **A** to **D** described below.
Briefly discuss the nature of the bonds between the metal atoms and also the reactions which interconvert the compounds.

1. Two moles of $Rh(\eta^5\text{-}Me_5Cp)(CO)_2$, **A**, react with one mole of ethereal $HBF_4$ with loss of one mole of CO to give **B**.
   **B** is a 1:1 electrolyte. The IR spectrum has bands at 2020 and $1820\ cm^{-1}$; it also has bands indicating the presence of the $[BF_4]^-$ anion. The $^1H$ NMR spectrum consists of a singlet at 2.03 ppm (30 H) and a 1:2:1 triplet at -10.4 ppm (1 H).

2. **B** is converted to a non-electrolyte **C** by the addition of a suitable base (such as $NaOCH_3$).
   **C** has relative molecular mass = $560\ g\ mol^{-1}$, IR bands at 1950 and $1780\ cm^{-1}$ and a $^1H$ NMR spectrum consisting of a singlet at 1.91 ppm.

3. When **C** is heated it loses one mole of CO to give **D**. **D** has one IR band in the $2100\text{--}1700\ cm^{-1}$ region at $1730\ cm^{-1}$. The mass spectrum shows a molecular ion $M^+_2$ at m/z = 532.

4. **A** can be converted directly to **D** by heating, with loss of one mole of CO per mole of **A**.

**Notes:**
- Only relevant IR bands are given;
- $^{103}Rh$ has I =   and is 100% abundant; and
- $Me_5Cp$ is the pentamethylcyclopentadienyl group.

**Prior knowledge**
Understanding of infrared, $^1H$ NMR spectroscopy and knowledge of stability.

**This problem is suitable for:**
- Second year students
- A tutorial – group work
- Approximately 45 minutes

**Knowledge/skills gained**
Practice in the use of infrared and $^1H$ NMR for structure determination.
Organometallic carbonyl chemistry.

# Identifying rhodium carbonyl compounds from chemical and spectroscopic data

**Solution**

1. $Rh(\eta^5\text{-}Me_5Cp)(CO)_2$, **A**, is an 18-electron molecule: no complications.

   **B** $^1$H NMR ratio of 30:1 suggests 2 $Me_5Cp$ groups and another single proton. The chemical shift of -10.4 ppm is typical of metal-bonded hydrides. The 1:2:1 triplet indicates that the hydrogen atom is symmetrically placed with respect to 2Rh atoms ($I = \frac{1}{2}$): *ie* μ-H.
   IR bands at 2020 and 1820 cm$^{-1}$ suggest a terminal CO.
   The loss of one mole of CO from two moles of **A** indicates overall stoichiometry of 2Rh:3CO:2Me$_5$Cp:1H
   **B** is:

   Note possible *cis-trans* isomers.

2. The base is likely to remove an acidic proton: the metal-bonded hydrogen atom has gone (NMR). IR frequencies are lower due to loss of positive charge.
   **C** is:

   Relative molecular mass and 18-electron rule are both consistent with this structure.

3. & 4. Relative molecular mass fits loss of CO from **C**; but IR indicates CO bridging only.

Possibly:

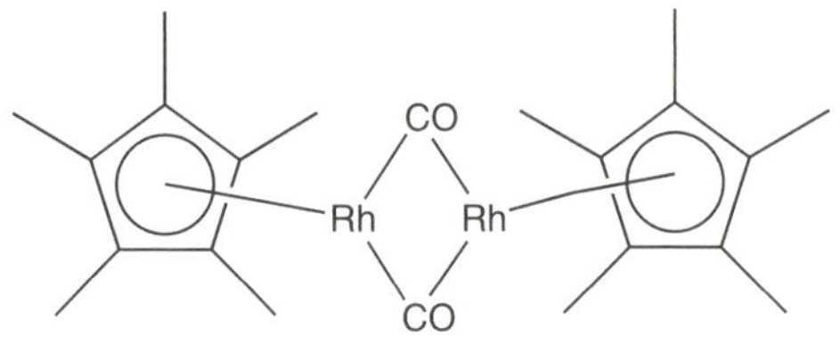

Relative molecular mass consistent but 18-electron rule is not. There are only 17 electrons, hence a double bond between Rh atoms.

**D** is:

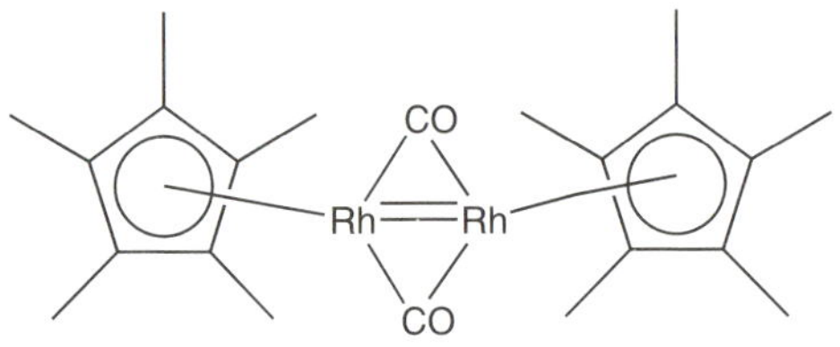

**Acknowledgement**
Chemistry Department, Glasgow University

## S34

# Identifying ruthenium compounds from chemical and spectroscopic data

**Problem**

A methanolic solution of ruthenium trichloride reacts with carbon monoxide resulting in a solution from which a white solid **A** can be obtained. The infrared spectrum of **A** shows a complex series of peaks all with frequencies between 2000 and 2200 $cm^{-1}$.

The white solid **A** reacts with thallium(I) cyclopentadienide to give **B** and with sodium cyclopentadienide to give **C**. If air is passed through a chloroform-ethanol solution of **C** containing hydrochloric acid, **B** is formed. The infrared spectrum of **C** shows a series of peaks above 2000 $cm^{-1}$ and a strong absorption band at 1794 $cm^{-1}$.

Analytical and molar mass figures obtained for the compounds **A**–**C** are:

Compound **A**: C, 14.10; H, 0.00; Cl, 27.70; O, 18.80; Ru, 39.90 %
Molar mass 512.0 g $mol^{-1}$

Compound **B**: C, 32.50; H, 1.93; Cl, 13.80; O, 12.40; Ru, 39.20 %
Molar mass 257.5 g $mol^{-1}$

Compound **C**: C, 37.90; H, 2.25; Cl, 0.00; O, 14.20; Ru, 45.60 %
Molar mass 444.0 g $mol^{-1}$

Suggest structures for the compounds **A**, **B** and **C**.
Suggest how **B** would react with sodium tetracarbonylcobaltate, $NaCo(CO)_4$.

**Prior knowledge**
Ability to recognise typical infrared bands; understanding of stabilities.

**This problem is suitable for:**
■ Second year students
■ A tutorial – group work
■ Approximately 40 minutes

**Knowledge/skills gained**
Practice with identification of compounds using infrared spectroscopy and working out empirical formulae.

# Identifying ruthenium compounds from chemical and spectroscopic data

**Solution**

**A** – No IR band near 1800 cm$^{-1}$, therefore no ($\mu$-CO).

Using the composition data:

| | | | |
|---|---|---|---|
| C | = | 14.10/12.00 | = 1.18 |
| H | = | 0.00 | |
| Cl | = | 27.70/35.50 | = 0.78 |
| Ru | = | 39.90/101.00 | = 0.40 |
| O | = | 18.80/16.00 | = 1.18 |

Divide by the smallest number:

| | | | |
|---|---|---|---|
| C | = | 1.18/0.40 | = 2.95 |
| Cl | = | 0.78/0.40 | = 1.95 |
| Ru | = | 0.40/0.40 | = 1.00 |
| O | = | 1.18/0.40 | = 2.95 |

By rounding to the nearest whole number we get:
C = 3, H = 0, Cl = 2, Ru = 1, O = 3

Therefore, the empirical formula is $RuCl_2(CO)_3$ giving:

This has 16 electrons.

In the empirical formula the ruthenium atom has sixteen electrons but for stability needs eighteen electrons. If the molecular formula is $Ru_2Cl_4(CO)_6$ and there are two chlorine bridges, the electron count rises to eighteen.

**B** is:

| | | | |
|---|---|---|---|
| C | = | 32.60/12.00 | = 2.72 |
| H | = | 1.90/1.00 | = 1.90 |
| Cl | = | 13.80/35.50 | = 0.39 |
| Ru | = | 39.20/101.00 | = 0.39 |
| O | = | 12.40/16.00 | = 0.78 |

Divide by the smallest number:

C  =  2.72/0.39      = 7.00
H  =  1.90/0.39      = 4.90
Cl =  0.39/0.39      = 1.00
Ru =  0.39/0.39      = 1.00
O  =  0.78/0.39      = 2.00

Giving C = 7, H = 5, Cl = 1, Ru = 1, O = 2

Therefore the structure of **B** is:

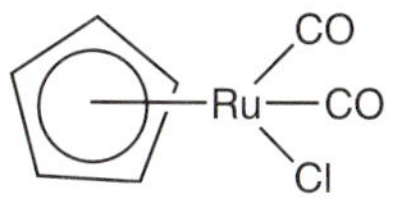

**C** is:

C  =  37.9/12.00     = 3.16
H  =  2.25/1.00      = 2.25
Cl =  0.00
Ru =  45.60/101      = 0.45
O  =  14.20/16.00    = 0.89

Divide by the smallest number:

C  =  3.16/0.45      = 7.02
H  =  2.25/0.45      = 5.00
Ru =  0.45/0.45      = 1.00
O  =  0.89/0.45      = 2.00

This gives C = 7, H = 5, Ru = 1, O = 2

**C** is $Ru_2(CO)_4(C_5H_5)_2$ with $\mu$–CO and terminal CO (from IR).

$CpRu(CO)_2$ is a 17-electron arrangement, therefore M-M bonded dimer likely.

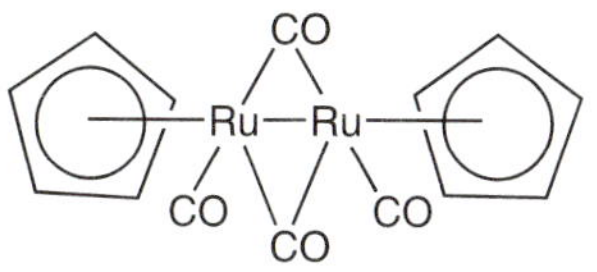

*Cis* isomer (a *trans* isomer also exists)

**B** is likely to react with $NaCo(CO)_4$ by NaCl elimination to yield:

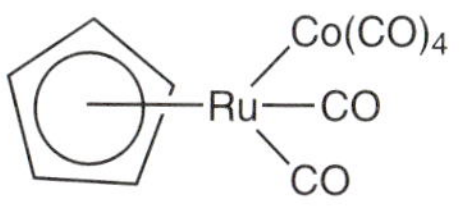

**Acknowledgement**
Chemistry Department, Glasgow University

# Identifying organic compounds from data

**Problem**

Consider the following reaction scheme:

$$A \xrightarrow{\text{Sodium amalgam}} B$$

$$B \xrightarrow{\text{Strong acid/water}} C \xrightarrow{\substack{\text{Potassium} \\ \text{hydrogensulfate}}} D$$

$$A \xrightarrow{\substack{\text{Ozonolysis or} \\ \text{(i) KMnO}_4 \\ \text{(ii) CrO}_3}} \text{Acetone + ethanedioic acid + 4-oxopentanoic acid}$$

$$A \xrightarrow{\substack{\text{Potassium} \\ \text{hydrogensulfate} \\ + \text{ heat}}} \text{4-methylisopropylbenzene}$$

$$D \xrightarrow{\text{Selenium/heat}} \text{4-methylisopropylbenzene}$$

1. Using the infrared spectra and the additional data provided below, identify compounds **A**, **B**, **C** and **D**; explain all your deductions.

2. Suggest a mechanism for the conversion of **B** into **C**.

3. Compound **A** is a well-known secondary metabolite. Using a one step retrosynthetic approach, analyse the structure of **A** and suggest a synthesis.

**Additional Data**

i)

| Molecular Formula | DBEs (calculate) | Pd/C + H$_2$ (room temp) No. moles H$_2$ taken up | $^1$H NMR No. of hydrogens (from integration) 4.5 to 7.5 ppm |
|---|---|---|---|
| **A** C$_{10}$H$_{16}$O | | 2 | 2 |
| **B** C$_{10}$H$_{18}$O | | 2 | 2 |
| **C** C$_{10}$H$_{18}$O | | 1 | 1 |
| **D** C$_{10}$H$_{16}$ | | 2 | 3 |

ii) Naturally occurring **C** is optically active and can be converted into **D** without loss of optical activity.

## Compound A

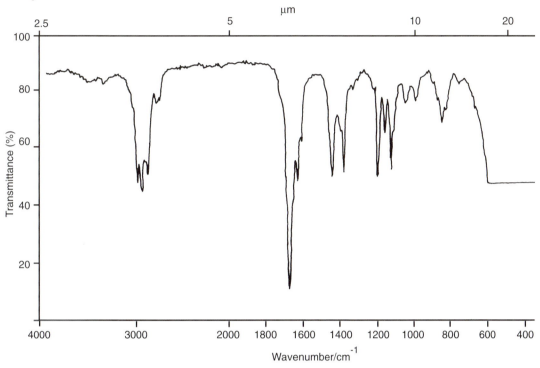

## Compound B

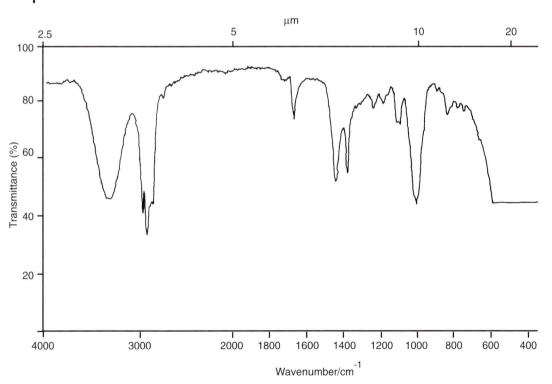

## Compound C

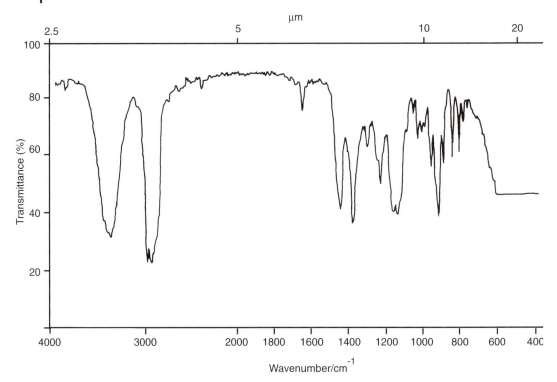

## Compound D

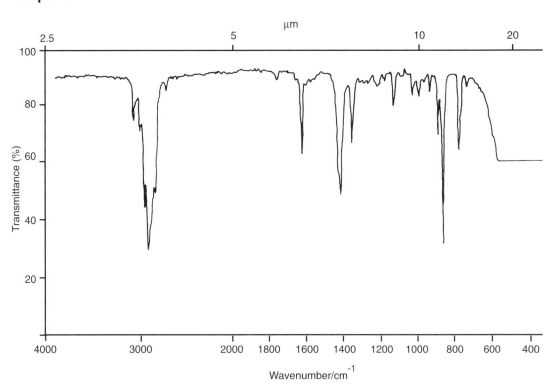

**Prior knowledge**

This problem is an exercise in infrared and [1]H NMR spectroscopy relating to
a series of natural products. Students require some knowledge in this area in
order to approach this problem successfully.

**This problem is suitable for:**
■ Second year students
■ A tutorial – group work
■ Approximately 2 hours

**Knowledge/skills gained**

Application and interpretation of infrared and [1]H NMR spectra to structure
elucidation.
The retrosynthetic approach required should give students new ideas on
solving reaction schemes.

# T35

# Identifying organic compounds from data

A lot of information is presented here, which must be digested and understood before the problem itself can be solved.

The problem can be solved in groups of approximately 4 people.

## Solution

1. **A** **B** **C** **D**

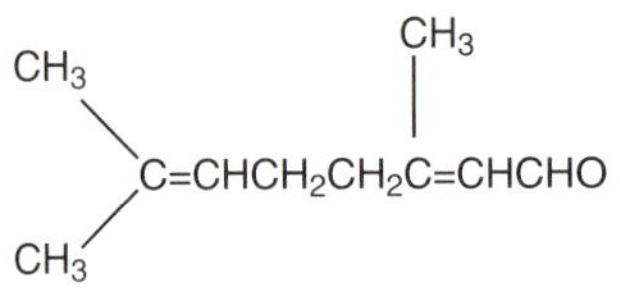

CH3
CH3
C=CHCH2CH2C=CHCHO
CH3


citral → geraniol → α-terpineol → limonene

**A**

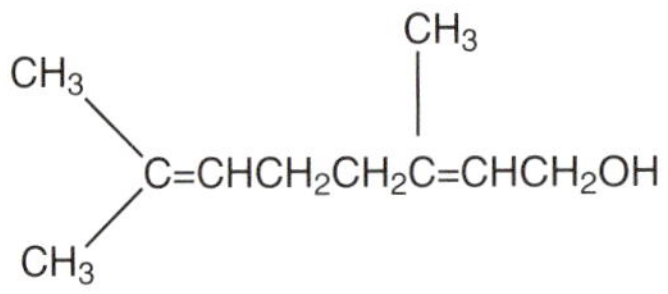

CH3
CH3
C=CHCH2CH2C=CHCH2OH
CH3


**B**

CH3
CH3
C
CH3
OH


**C**

CH3
CH3
C
CH2


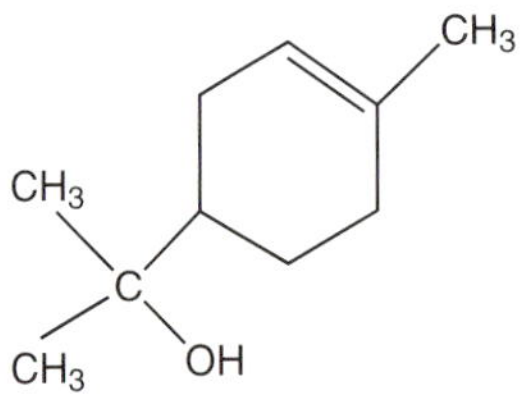

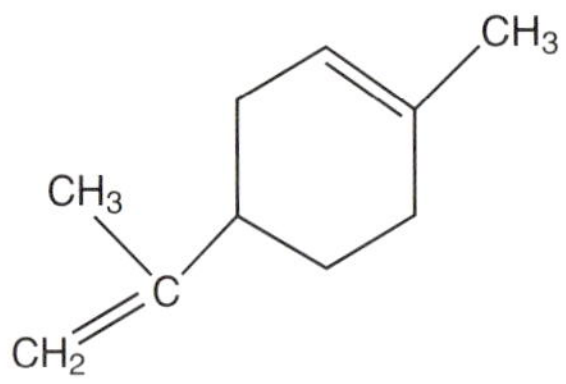

**D**

Analysis of infrared:

| | | |
|---|---|---|
| **A** | 1380 $cm^{-1}$ | C–H |
| | 1630 $cm^{-1}$ | C=C |
| | 1675 $cm^{-1}$ | C=O |
| | 2760 $cm^{-1}$ | C–H (aldehyde stretch) |
| | 2900 $cm^{-1}$ | C–H sat. |

| | | |
|---|---|---|
| **B** | 1000 $cm^{-1}$ | C–O |
| | 1380 $cm^{-1}$ | C–H |
| | 1675 $cm^{-1}$ | C=O |
| | 2900 $cm^{-1}$ | C–H sat. |
| | 3300 $cm^{-1}$ | OH |

| | | |
|---|---|---|
| **C** | 1150 $cm^{-1}$ | C–O |
| | 1380 $cm^{-1}$ | C–H |
| | 1630 $cm^{-1}$ | C=C |
| | 2900 $cm^{-1}$ | C–H sat. |
| | 3300 $cm^{-1}$ | OH |

| | | |
|---|---|---|
| **D** | 1380 $cm^{-1}$ | C–H |
| | 1630 $cm^{-1}$ | C=C |
| | 2900 $cm^{-1}$ | C–H sat. |
| | 3050 $cm^{-1}$ | C–H (alkene stretch) |

2.    Mechanism for the conversion of **B** to **C**:

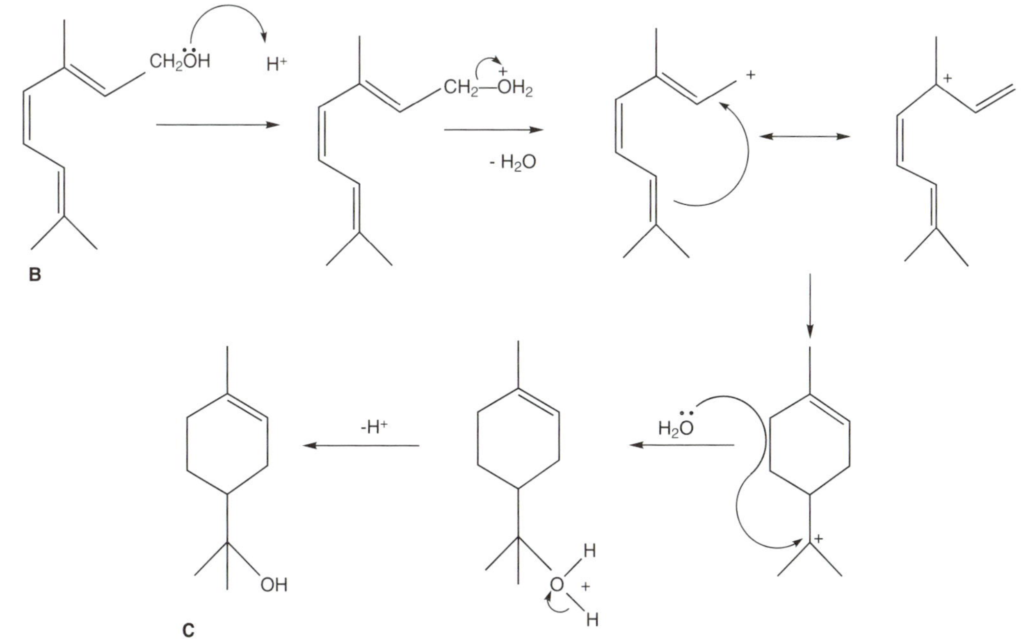

3.    Retrosynthetic analysis

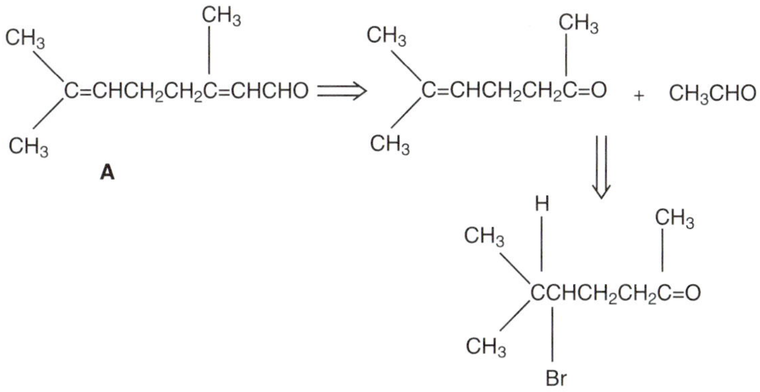

Suggested Synthesis of **A**

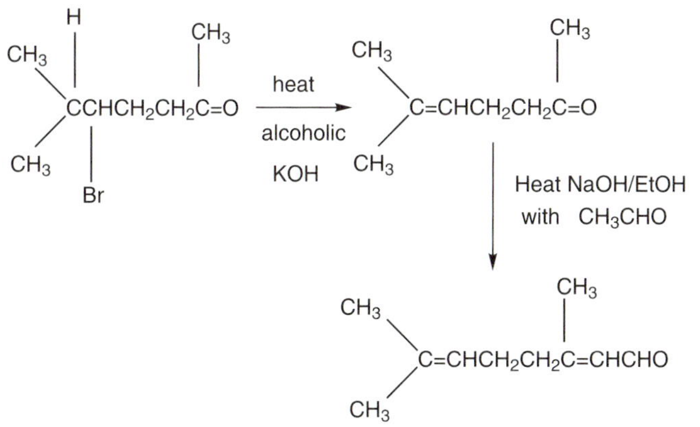

**Acknowledgement**

Dr Dave Bannister, Manchester Metropolitan University

# Monitoring iron removal from blood by UV/visible spectroscopy

**Problem**

Transferrin is an iron-transport protein found in blood. It has a molar mass of 81 000 g mol$^{-1}$ and carries two Fe(III) ions.

Desferrioxamine B is a potent iron chelator used to treat patients with iron overload. It has a molar mass of about 650 g mol$^{-1}$ and can bind one iron atom as Fe(III).

Desferrioxamine can take iron from many sites within the body and is excreted (with its iron) through the kidneys. The molar absorptivities of these compounds (saturated with iron) at two wavelengths are given below.

Both compounds are colourless in the absence of iron.

| Wavelength | $\varepsilon$ / dm$^3$ mol$^{-1}$ cm$^{-1}$ | |
| --- | --- | --- |
| nm | Transferrin | Desferrioxamine |
| 428 | 3540 | 2730 |
| 470 | 4170 | 2290 |

$$\lambda_{max} \text{ (Transferrin)} = 470 \text{ nm}, \quad \lambda_{max} \text{ (Desferrioxamine)} = 428 \text{ nm}$$

(a) A solution containing transferrin only, exhibits an absorbance of 0.46 at 470 nm in a 1.00 cm cell. Calculate the concentration of transferrin in milligrams per cm$^3$ and the concentration of iron in grams per dm$^3$.

(b) A short time after adding some desferrioxamine (which dilutes the sample), the absorbance at 470 nm was 0.42, and the absorbance at 428 nm was 0.40. Calculate the fraction of iron in transferrin which binds two iron atoms as a weight percentage and in desferrioxamine (which binds only one iron atom).

**Prior knowledge**
Beer-Lambert Law and ability to perform calculations.

**This problem is suitable for:**
■ Second year students
■ A tutorial – group or individual work
■ Approximately 45 minutes

**Knowledge/skills gained**
Developing skills in calculations involving Beer-Lambert Law, dilution factors and converting units.

## T36

# Monitoring iron removal from blood by UV/visible spectroscopy

This problem could be tackled as a group or individually. Reference books may be required if students cannot recall the Beer-Lambert Law and its definitions.

**Solution**

(a) The transferrin (T) is only coloured if it contains iron, so the absorbance of the solution containing only T gives the total amount of iron started with, which is just redistributed on adding the desferrioxamine (D).

According to Beer-Lambert Law

| | A | $= \varepsilon c l$ |
|---|---|---|
| Where | A | = absorbance |
| | $\varepsilon$ | = molar absorptivity |
| | c | = concentration |
| | l | = path length |

| Therefore, | 0.46 | $= 4170 \text{ dm}^3 \text{ mol}^{-1} \text{ cm}^{-1} \times c \times 1.00 \text{ cm}$ |
|---|---|---|
| | c | $= 0.46/4170 \text{ mol dm}^{-3}$ |
| | | $= 1.10 \times 10^{-4} \text{ mol dm}^{-3}$ |
| Molar mass | | $= 81\ 000 \text{ g mol}^{-1}$ |

Therefore, for $1.10 \times 10^{-4} \text{ mol dm}^{-3}$

| concentration | $= (1.10 \times 10^{-4} \text{ mol dm}^{-3}) \times (81\ 000 \text{ g mol}^{-1})$ |
|---|---|
| | $= 8.91 \text{ g dm}^{-3}$ |
| | $= 8.91 \text{ mg cm}^{-3}$ |

| concentration of iron | $= 2 (1.10 \times 10^{-4}) \text{ mol dm}^{-3}$ |
|---|---|
| Molar mass | $= 55.85 \text{ g mol}^{-1}$ |

Therefore,

| concentration of iron | $= 55.85 \text{ g mol}^{-1} (2.20 \times 10^{-4} \text{ mol dm}^{-3})$ |
|---|---|
| | $= 1.23 \times 10^{-2} \text{ g dm}^{-3}$ |

(b) We have two simultaneous linear equations, dictated by the fact that T and D both absorb at 428 nm and 470 nm. Thus:

| (I) | 0.42 | $= 4170T + 2290D$ |
|---|---|---|
| (II) | 0.40 | $= 3540T + 2730D$ |
| From (I) | D | $= 0.42 - 4170T/2290$ |

substitute into (II)

$$0.40 = 3540T + 2730/2290\ (0.42 - 170T)$$
$$0.40 = 3540T + 0.50 - 4971T$$
$$1431T = 0.10$$

Therefore,

$$T = 6.99 \times 10^{-5}\ mol\ dm^{-3}$$
$$D = 5.61 \times 10^{-5}\ mol\ dm^{-3}$$

Concentration of Fe in D $= 5.61 \times 10^{-5}\ mol\ dm^{-3}$
Concentration of Fe in T $= 2 \times 6.99 \times 10^{-5}\ mol\ dm^{-3}$

Therefore,

% Fe in D $= 5.61/14 + 5.61 \times 100$
$= 28.61\%$
% Fe in T $= 14/14 + 5.61 \times 100$
$= 71.39\%$

**Acknowledgement**

Dr John Smith, University of Newcastle upon Tyne

## S37

# UV/visible spectroscopy of a zinc porphyrin

**Problem**

A zinc porphyrin, **X**, was dissolved in an organic solvent and the absorbance measured at 435 nm as a function of the concentration of **X** in a 0.1 cm path length cell.

The following results were obtained:

| $[\mathbf{X}]/10^{-6}$ mol dm$^{-3}$ | 1.00 | 3.20 | 7.40 | 9.10 | 12.00 | 16.00 |
|---|---|---|---|---|---|---|
| Absorbance/10$^{-2}$ | 1.80 | 3.60 | 6.90 | 8.30 | 10.60 | 13.80 |

The reference employed was an identical empty cell.

(a) What is the extinction coefficient of the zinc porphyrin solution at 435 nm?

(b) Given that the concentration of the pure solvent was 19.20 mol dm$^{-3}$, what was the molar extinction coefficient of the solvent at 435 nm?

(c) If the reference cell was filled with the organic solvent, and the experiment repeated, how would you expect the results to change?

(d) Using the pure solvent in the reference cell, what concentration of zinc porphyrin would result in (i) 33% and (ii) 1% of the light incident on the sample cell being transmitted?

**Prior knowledge**
Knowledge of the Beer-Lambert Law is needed and ability to draw graphs; working out gradient and intercept.

**This problem is suitable for:**
- ■ Second year students
- ■ A tutorial – group work
- ■ Approximately 45 minutes

**Knowledge/skills gained**
Practice in absorbance calculations and application of the Beer-Lambert Law.

# UV/visible spectroscopy of a zinc porphyrin

Even though the compound used in this problem is zinc porphyrin it is not necessary that porphyrins should have been studied.

Porphyrins are found as metal complexes in several enzymes.

**Solution**

(a) This question should be answered by plotting a graph of the data absorbance, A versus [**X**].

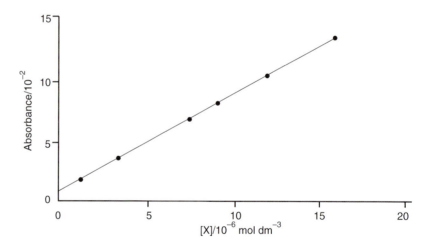

From the graph the slope can be calculated to be 8000 $mol^{-1}$ $dm^3$ = $\varepsilon l$.

Since $l$ = 0.10 cm then $\varepsilon$ = 8.00 x $10^4$ $dm^3$ $mol^{-1}$ $cm^{-1}$.

(b) Again, using the graph plotted in part (a) the intercept on the Absorbance axis is 0.01 and the question can now be answered using Beer-Lambert Law.

$\varepsilon$ = A/cl
$\varepsilon$ = 0.01/19.20 x 0.10
$\varepsilon$ = 5.21 x $10^{-3}$ $dm^3$ $mol^{-1}$ $cm^{-1}$

(c) No intercept, the line would pass through the origin. The absorbance will fall.

(d) absorbance, $\quad A = \log (I_0/I)$
% transmittance, $T = (I/I_0) \times 100\%$

If $T = 33\%$ then $A = 0.48$
If $T = 1\%$ then $\quad A = 2.00$

$$A = \varepsilon cl$$
$$c = A/\varepsilon l$$
$$A = 0.48 \text{ (or 2.00)}$$
$$\varepsilon = 80\ 000\ \text{dm}^3\ \text{mol}^{-1}\ \text{cm}^{-1}$$
$$l = 0.1\ \text{cm}$$

Therefore,

(i) $\quad c = 6.00 \times 10^{-5}\ \text{mol dm}^{-3}$
(ii) $\quad c = 2.50 \times 10^{-4}\ \text{mol dm}^{-3}$

**Acknowledgement**

Dr John Smith, University of Newcastle upon Tyne

S38

# Quantitative infrared analysis of sebum extract

**Problem**

A student's forehead was wiped with a sterile cotton wool ball. The cotton wool ball was extracted using a Soxhlet apparatus with 100 cm$^3$ trichloromethane. The trichloromethane solvent was removed from the extract using a rotary evaporator to leave a sebum residue of 4.15 x 10$^{-2}$ g. The sebum residue was redissolved in 6 cm$^3$ of trichloromethane. Approximately 0.1 cm$^3$ of this solution was transferred to a NaCl liquid cell (path length = 0.43 mm) using a pipette. The percent transmittance values were obtained for the ester and acid carbonyl stretches. (Note: the baseline for each peak was 100% transmittance).

Palmitic acid = 40.49%T          Glycerol trioleate (ester) = 58.13%T

A series of calibration solutions in trichloromethane was produced and their peak percent transmittance values obtained for the carboxylic acid (C$_{15}$H$_{31}$COOH) and ester carbonyl (C$_{17}$H$_{33}$COO-CH$_2$-CH(-OOCC$_{17}$H$_{33}$)-CH$_2$-OOCC$_{17}$H$_{33}$) stretches using a NaCl liquid cell (path length = 0.45 mm).

| Palmitic acid | | Glycerol trioleate | |
|---|---|---|---|
| Concentration/ (mol dm$^{-3}$ x 10$^{-3}$) | %T | Concentration/ (mol dm$^{-3}$ x 10$^{-3}$) | %T |
| 3.90 | 73.20 | 2.26 | 70.97 |
| 7.81 | 53.58 | 4.53 | 50.37 |
| 15.63 | 28.71 | 6.79 | 35.74 |
| 23.44 | 15.39 | 9.10 | 25.37 |
| 31.25 | 8.24 | 11.31 | 18.00 |
| 39.06 | 4.42 | 13.58 | 12.78 |

1. Determine the % (mass) level of ester and acid in the sebum extract.

2. State any assumptions used to obtain the answers.

**Prior knowledge**
Beer-Lambert Law, calibration graphs and calculations.

**This problem is suitable for:**
■ Second year students
■ A tutorial – group work
■ Approximately 1 hour

**Knowledge/skills gained**
Application of quantitative results from infrared analysis.

# Quantitative infrared analysis of sebum extract

**Solution**

1.  Convert %T to A using $(A = -\log(T/T_o))$
    where T = % transmitted radiation; $T_o$ = 100%

    Molar mass for palmitic acid is 256 g $mol^{-1}$
    and for glycerol trioleate is 884 g $mol^{-1}$

    Convert mol $dm^{-3}$ to g $dm^{-3}$ – this doesn't have to be done but will make plotting the graph easier.

| Palmitic acid | | Glycerol trioleate | |
|---|---|---|---|
| Concentration/ (g $dm^{-3}$) | %T | Concentration/ (g $dm^{-3}$) | %T |
| 1 | 0.14 | 2 | 0.15 |
| 2 | 0.27 | 4 | 0.30 |
| 4 | 0.54 | 6 | 0.45 |
| 6 | 0.81 | 8 | 0.60 |
| 8 | 1.08 | 10 | 0.74 |
| 10 | 1.35 | 12 | 0.89 |

Acid  Y = 0.14X                    Ester Y = 0.07X

Calculate the molar absorptivity from each calibration graph.

For the acid:
1 A unit is equivalent to 7.38 g $dm^{-3}$ (obtained from gradient)

using the Beer-Lambert Law
A     = $\varepsilon$cl
$\varepsilon$     = molar absorptivity /($dm^3$ $mol^{-1}$ $cm^{-1}$)
c     = concentration /(mol $dm^{-3}$)
l     = pathlength /(cm)

Therefore,
$\varepsilon$     = A/cl
$\varepsilon_{acid}$ = 770.75  $dm^3$ $mol^{-1}$ $cm^{-1}$

For the ester:
1 A unit is equivalent to 13.43 g $dm^{-3}$ (obtained from gradient)
$\varepsilon_{ester}$ = 1462.84  $dm^3$ $mol^{-1}$ $cm^{-1}$

Using the $\varepsilon$ values the unknown concentrations can be calculated

Acid

| | |
|---|---|
| A | = 0.39 |
| conc/(mol dm$^{-3}$) | = 0.39/(770.75 x 0.04) |
| | = 1.27 x 10$^{-2}$ |
| conc/(g dm$^{-3}$) | = 3.24 |
| mass acid | = 3.24 x 6 (cm$^3$ CHCl$_3$) |
| | = 1.94 x 10$^{-2}$ g |
| % acid (by mass) | = (1.94 x 10$^{-2}$/4.15 x 10$^{-2}$ ) x 100 |
| | = 44% by mass |

Ester

| | |
|---|---|
| A | = 0.24 |
| conc/(mol dm$^{-3}$) | = 0.24/(1462.84 x 0.24) |
| | = 4.10 x 10$^{-3}$ |
| conc/(g dm$^{-3}$) | = 3.63 |
| mass ester | = 3.63 x 6 (cm$^3$ CHCl$_3$) |
| | = 2.18 x 10$^{-2}$ g |
| % ester (by mass) | = (2.18 x 10$^{-2}$/4.15 x 10$^{-2}$) x 100 |
| | = 52% by mass |

2.  Assumptions:

    (i)  The Beer-Lambert Law is only linear up to 1.00 A unit (greater than 1 on
         FTIR instruments). The absorbance values are well within this limit.

    (ii) There are no ester/acid groups in the cotton wool (otherwise it would be
         necessary to run a blank).

    (iii) Peak interferences with each other are negligible. In the example given
          however, because the absorbance intensities for each peak are similar
          there will be an error of approximately +/-3%.

### Acknowledgement
Dr Jeremy Andrew and Dr Allen Millichope, Unilever Research, Port
Sunlight, Wirral

## S39

# Qualitative infrared analysis of a shower gel extract

**Problem**

A shower gel was extracted with 3 cm$^3$ ethoxyethane and 3 cm$^3$ ethanol. The soluble fractions were examined by casting on to a KBr plate.

The ethoxyethane fraction indicated peaks at the following positions: 2923, 2855, 1464, 1223, 1117, 1027, 930, 587 cm$^{-1}$.

The ethanol fraction gave peaks at the following positions: 2919, 2850, 1558, 1468 cm$^{-1}$.

Using the characteristic assignment table, identify the peaks and classify the compounds.

**Prior knowledge**
Ability to classify compounds using information given.
Infrared spectroscopy.

**This problem is suitable for:**
∎ Second year students
∎ A tutorial – group work
∎ Approximately 1 hour

**Knowledge/skills gained**
Revision in the identification of compounds using data obtained from infrared spectroscopy.

**Characteristic group frequencies**

| Group | Wavenumber range (cm$^{-1}$) | Group | Wavenumber range (cm$^{-1}$) |
|---|---|---|---|
| $-O-SO_3^-$ bend | 620–560 | $COO^-$ carboxylate stretch | 1600–1550 |
| $-CH_2$ rock | 740–720 | $-NH$ bend | 1655–1515 |
| C–Cl stretch | 800–600 | C=C stretch | 1680–1510 |
| $-CH_2CH_2O-$ stretch | 980–930 | C=O acid stretch | 1705–1680 |
| $-C-OH$ primary alcohol stretch | 1050–1000 | C=O ketone stretch | 1720–1705 |
| $-O-SO_3^-$ symmetric stretch | 1080–1020 | C=O aldehyde stretch | 1720–1710 |
| C–OH secondary alcohol stretch | 1100–1050 | C=O ester stretch | 1780–1740 |
| $-C-O-C$ ether stretch | 1160–1100 | $-CH_2$ symmetric stretch | 2850–2820 |
| C–OH tertiary alcohol stretch | 1200–1100 | $-CH_2$ asymmetric stretch | 2980–2910 |
| $-O-SO_3^-$ asymmetric stretch | 1230–1180 | H-bonded OH stretch | 3000–2500 |
| $-CH_3$ umbrella | 1385–1365 | =C–H stretch | 3150–3000 |
| $-CH_2$ bend | 1470–1400 | $-NH$ stretch | 3520–3100 |
| $-C-N$ stretch | 1400–920 | $-OH$ stretch | 3650–3200 |

**Common detergent components**

| | |
|---|---|
| Sodium alkyl benzene sulfonate | $R-Ar-SO_3^-\ Na^+$ |
| Alkyl ethoxylate | $R-(CH_2CH_2O-)_nH$ |
| Sodium alkyl ether sulfate | $R-(CH_2CH_2O-)_nSO_3^-\ Na^+$ |
| Soap (long alkyl chain carboxylate) | $R-COO^-\ Na^+$ |
| Fatty acid (alkanoic acid) | $R-COOH$ |
| Glycerol (propane-1,2,3-triol) | $HO-CH_2-CH(OH)-CH_2-OH$ |
| Dimethyl siloxane | $CH_3-(Si(CH_3)_2O_2)_n-CH_3$ |
| Sodium alkyl sulfate | $R-O-SO_3^-\ Na^+$ |
| Sodium alkyl sulfonate | $R-SO_3^-\ Na^+$ |

### Ethoxyethane soluble spectrum

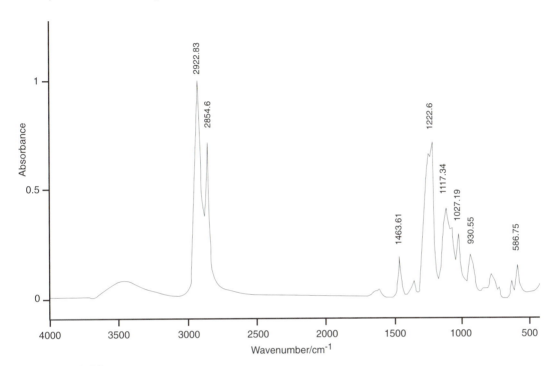

### Ethanol soluble spectrum

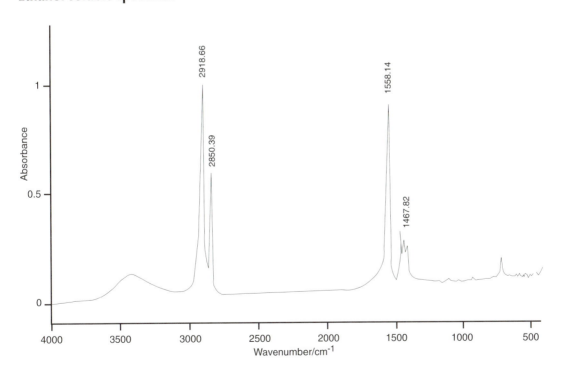

# Qualitative infrared analysis of a shower gel extract

**Solution**

### Ethoxyethane solution

Peak positions

| | | |
|---|---|---|
| 2923 | $-CH_2$ | asymmetric stretch |
| 2855 | $-CH_2$ | symmetric stretch |
| 1464 | $-CH_2$ | bend |
| 1223 | $-O-SO_3^-$ | asymmetric stretch |
| 1117 | $-C-O-C$ | ether stretch |
| 1027 | $-O-SO_3^-$ | symmetric stretch |
| 930 | $-CH_2CH_2O-$ | stretch |
| 587 | $-O-SO_3^=$ | bend |

Compound:
alkyl ether sulfate

### Ethanol solution

Peak positions

| | | |
|---|---|---|
| 2919 | $-CH_2$ | asymmetric stretch |
| 2850 | $-CH_2$ | symmetric stretch |
| 1558 | $COO^-$ | carboxylate stretch |
| 1468 | $-CH_2$ | bend |

Compound:
soap

### Acknowledgement
Dr Jeremy Andrew and Dr Allen Millichope, Unilever Research, Port Sunlight, Wirral

## S40

# Identifying ZCl₄ from chemical and infrared data

**Problem**

When element **Z** reacts with chlorine it gives a compound of empirical formula **Z**Cl$_4$ with a boiling point of 57 °C.

The infrared spectrum of the liquid has bands at 608 and 221 cm$^{-1}$ whilst the Raman spectrum has bands at 424 (polarised), 608, 221 and 150 cm$^{-1}$ (depolarised).

The chloride fumes in moist air and reacts readily in water giving a solution which on evaporation yields first a colourless gel and, after more intense drying, a colourless, hard, amorphous solid which though insoluble in water can reversibly absorb quite large quantities of it.

Identify element **Z** as far as possible and account for all the above observations. How could you be sure of your identification?

**Prior knowledge**
Raman and infrared spectroscopy.

**This problem is suitable for:**
■   Second year students
■   A tutorial – group work
■   Approximately 1 hour

**Knowledge/skills gained**
Ability to put together basic concepts on IR and Raman spectroscopy to solve a problem for an unknown compound.

T40

# Identifying ZCl$_4$ from chemical and infrared data

An outline solution is provided below.

**Solution**

The halide is volatile, and is therefore likely to be covalent.

If the formula is **Z**Cl$_4$ then 9 vibrations are expected. IR and Raman spectra show only 4 distinct bands so some are likely to be degenerate – hence the molecule will be fairly symmetrical (at least a threefold axis).

Bands are at the same frequencies in IR and Raman spectra, so **Z**Cl$_4$ is not centrosymmetric. Possible shapes are: square pyramidal or tetrahedral (high symmetry – so probably this).

Tetrahedral tetrachlorides are formed by group IV elements: C, Si, Ge, Sn, Pb, Ti, Zr, Hf.

As we are told the chloride readily reacts with water, it cannot be CCl$_4$.

The gel is a hydroxide that on drying goes to oxide. The oxide is amorphous so it does not have an ionic lattice. It reversibly absorbs water. This suggests covalent bonding – ionic oxides would 'anneal' and give compact, non-reactive structures. Non-directional ionic bonds allow the structure to accommodate water more readily than the directional covalent bonds that must be broken, giving a high activation energy for rearrangement.

Thus **Z** is probably silicon (giving silica gel) but might be germanium. The exact vibrational frequencies would confirm this assumption.

**Acknowledgement**
Dr Colin Peacock, Lancaster University

**S41**

# Identify ACl$_4$ from chemical and infrared data

**Problem**

Element **A** reacts readily with chlorine at room temperature to give a solid product with a stoichiometry of four chlorine atoms to one atom of **A**. This colourless material sublimes at 191 °C and is considerably dissociated into a dichloride and chlorine to give a reddish vapour.

The infrared spectrum of the solid has bands at
378, 347, 275, 206, 190, 165 and 145 cm$^{-1}$,

whilst the Raman spectrum has bands at
378 (polarised), 347, 206, 165 and 127 cm$^{-1}$.

The solid may be dissolved in concentrated aqueous hydrochloric acid and following the addition of KCl, a salt analysed as **A**K$_2$Cl$_6$ can be isolated. The IR spectrum of this salt shows two bands only at 275 and 153 cm$^{-1}$ and the Raman spectrum three bands: 290 (polarised), 255 and 165 cm$^{-1}$.

Identify element **A** as far as possible and account for the above observations.

**Prior knowledge**
This problem deals with main group inorganic chemistry. Knowledge is required on infrared and Raman spectroscopy and shapes of inorganic compounds.

**This problem is suitable for:**
■ Second year students
■ A tutorial – group work
■ Approximately 1 hour

**Knowledge/skills gained**
The problem encompasses spectroscopy and inorganic chemistry. A greater understanding of how these techniques can be used to identify elements should be achieved through this problem.

T41

# Identify $ACl_4$ from chemical and infrared data

This problem could be solved in small groups. The solution suggested below involves a process of eliminating factors in order to find the answer.

**Solution**

The stoichiometry: four chlorine atoms to one atom of **A** gives a tetrachloride. A sublimation temperature of 191 °C suggests that the molecule is associated to quite a large extent. It is more likely to be a chloride of one of the less electronegative elements, even if not fully ionic.

The vibrational spectrum has 8 distinct frequencies:
378, 347, 275, 206, 190, 165, 145, 127 $cm^{-1}$, so the molecule is probably monomeric (5 atoms, 9 vibrations).

IR and Raman bands coincide, so not centrosymmetric. Eight of the nine vibrations are active, so not highly symmetrical.

Possible shapes for structure:

| tetrahedral | square planar | distorted square pyramid |
|---|---|---|
| (ruled out by high symmetry) | (ruled out as centrosymmetric) | (probably this) |

Valence Shell Electron Pair Repulsion Theory (VSEPR) predicts that this arrangement would be given by a structure with four chorine atoms and five electron pairs, hence from group VI of the periodic table *ie* S, Se or Te.

The vibrational spectra of the adduct with KCl show high symmetry and a centrosymmetric shape. So $ACl_6^{2-}$ is octahedral with $K^+$ counterions. This certainly rules out sulfur which is too small for six chlorines to pack around it.

Hence element **A** is likely to be selenium. Tellurium is a possibility, however, it could be ruled out using knowledge of approximate boiling points of semi-ionic compounds.

Extra marks could be awarded to students who attempt to explain the colour in terms of presence or absence of low-lying molecular orbitals; ease of reaction from electronegativity differences; dissociation of tetrachloride in gas phase through loss of lattice energy which stabilises the solid tetrachloride; lack of hydrolysis in concentrated aqueous HCl from the ionic nature of the A-Cl bond.

**Acknowledgement**
Dr Colin Peacock, Lancaster University

## S42

# Determination of the solubility of calcium hydroxide

**Problem**

You have been given a saturated solution of calcium hydroxide in water at 298 K.

Design an experiment that will enable the solubility of the calcium hydroxide in the above solution to be determined.

Your report should include:
■ an outline of your approach;
■ the sources and values of any data from the literature; and
■ full explanations of equations and assumptions.

**Prior knowledge**
Communication skills.

**This problem is suitable for:**
■ First year students
■ A tutorial – group work (three or four students per group)
■ Pre-laboratory session
■ Approximately 2 hours

**Knowledge/skills gained**
Improved skills in communication and working with others; knowledge of solubility.

# Determination of the solubility of calcium hydroxide

The following questions are supplied to prompt students.

1. How is solubility measured?

2. What titrant is to be used?

3. What will the concentration of the titrant be?

4. What volume of sample will be taken?

5. What indicator will be used?

6. What is the solubility and the solubility product constant of calcium hydroxide at 298 K?

The students may require the use of textbooks to answer some of these questions. The students should focus on a method in which the concentration of a saturated solution is determined by titration with a suitable reagent (*eg* HCl or EDTA).

**Solution**

A possible method is suggested.

Solubility of $Ca(OH)_2 = [Ca^{2+}]$
If we can find the concentration of calcium ions we will know the solubility of the calcium hydroxide.

The method suggested is a titration with EDTA.

Measure 25 cm$^3$ of the saturated calcium hydroxide solution into a 250 cm$^3$ conical flask, dilute the solution with 25 cm$^3$ distilled water and add 2 cm$^3$ of a basic buffer solution (ammonia solution with ammonium chloride), 1 cm$^3$ Mg–EDTA (0.1 mol dm$^{-3}$) and approximately 35 mg eriochrome black/ potassium nitrate indicator. Titrate the solution with 0.1 mol dm$^{-3}$ EDTA solution. The colour changes from red to blue at the endpoint. Repeat this procedure several times to achieve consistent results. Note down the volume of EDTA solution required and using this information calculate the concentration of the calcium ions and hence the solubility of calcium hydroxide.

**Acknowledgement**
Dr Jim Wood, University of Huddersfield

## S43

# Water analysis of a stream – planning the approach

**Problem**

You have been given the job of routine analysis of a local stream.

Discuss your method for this sampling procedure including (among other topics):
■ frequency;
■ storage;
■ number; and
■ location.

**Prior knowledge**
Sampling.

**This problem is suitable for:**
■ First year students
■ A tutorial – group work
■ Approximately 45 minutes

**Knowledge/skills gained**
A greater understanding of sampling strategies and brainstorming skills.

# Water analysis of a stream – planning the approach

This is an open-ended problem and is most suitable for brainstorming in tutorials. Sampling is a popular area in chemistry and many students will go on to work in a related field.

As indicated in the problem, frequency, storage, number and location are some of the issues to be discussed. The students should appreciate that all these matters have to be considered in theory before any practical on-site sampling can take place.

A major factor for consideration is what is being analysed and what techniques are required. This is the key issue in sampling.

A good approach to this could be brainstorming in the form of a web diagram with key issues scribbled down as they come to mind and then being elaborated upon.

**Solution**

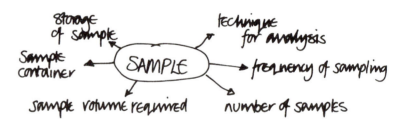

While expanding ideas, new ideas may evolve and can easily be written down and slotted into the web diagram.

**Frequency of sampling** – this could be affected by time and cost and also by the known frequency of contaminant release; day and night may also be a consideration for concentration of contaminant.

**Sample container** – the volatility of the sample should be considered here (contamination from the container).

**Storage of sample** – light sensitivity, is it likely to undergo microbial degradation?

Some other things that could be mentioned are:
■ Representative samples should be taken at regular intervals along the stream to give a systematic and fair result.

■ Are there any industries around that may be using the stream to cool machinery or releasing contaminants?

■ Is the stream populated? If so, what creatures live there and are these creatures known to thrive in any particular type of environment?

■ Does the vegetation surrounding the stream look healthy?

# An analytical chemist – steps in the analysis of a water sample

**Problem**

Imagine you are a qualified chemist and are working as an environmental analyst. You have been given a water sample to analyse for possible contamination. What would be your first thoughts?

**Prior knowledge**
This problem mainly requires chemical common sense and brainstorming.

**This problem is suitable for:**
■ First year students
■ A tutorial – group work or class discussion
■ Approximately 30 minutes

**Knowledge/skills gained**
Analytical methods.

# An analytical chemist – steps in the analysis of a water sample

This problem could be used at the beginning of a series of lectures on analysis.

The question is designed to allow students plenty of scope for thinking and brainstorming. Either in small groups or as a class allow the students to brainstorm for 5–10 minutes without any assistance.

## Solution

Initial thoughts when given the sample:

- Where did the sample come from?
- Who took the sample?
- What information have I been given about the sample?
- What can I find out about the area where the sample was taken?
- What am I looking for in the sample?
- Have I done anything like this before?
- If so, how did I approach it?
- Could I use the same method?
- Did I make notes last time?
- Do I know of anyone else who has done something similar?
- Is the sample likely to be dangerous?
- What precautions should I take when performing the analysis?

You should then consider any specifications you may have been given:

- How much time do I have?
- How much do I have to spend?
- What instruments do I have available for the analysis?
- How much sample do I have?

Once the above have been decided, you can then determine whether the sample needs pretreatment before analysis by your chosen method.

# A silvery-white element – identification and uses

**Problem**

You are given a sample of an element. It is a silvery-white metal.

Identify and suggest uses for this element, given the following information:

melting point     = 1802 K
boiling point      = 3136 K

The element exhibits low toxicity.

**Prior knowledge**
The basic properties and characteristics of elements.

**This problem is suitable for:**
■ Second year students
■ A tutorial – group work or class discussion
■ Approximately 45 minutes

**Knowledge/skills gained**
Communication and information retrieval skills.

T45

# A silvery-white element – identification and uses

This problem could be used for any silvery-white element.

The problem could be extended by incorporating more analytical techniques to gain a more accurate assessment.

A suggested solution is outlined below.

**Solution**

From the data a number of elements can be eliminated. For example, gases and liquids (at room temperature), gold and copper.

The following elements are silvery/silvery-white metals:

**Actinium** – boiling point similar, silvery-white metal, toxic due to radioactivity.
**Cobalt** – melting and boiling points similar, silvery-blue metal, carcinogenic.
**Erbium** – melting and boiling points match, silvery-white metal, low toxicity.
**Iron** – melting and boiling point similar, silvery metal, carcinogenic.
**Palladium** – melting point similar, silvery-white metal, non-toxic generally.
**Scandium** – melting and boiling points similar, soft silvery-white metal, suspected carcinogen.
**Thallium** – melting point similar, silvery metal, low toxicity.
**Yttrium** – melting point similar, silvery-white metal, suspected carcinogen.

These elements show similarities to some or all of the properties listed in the problem. Erbium is the only element exhibiting all the properties.

Suggested uses of erbium:
■ Nuclear reactor hardware – due to the metal's high cross-section for the absorption of thermal neutrons.
■ Display panels and colour television picture tubes.
■ Lasers and semiconductors.

## S46

# Orange sublimate – an exercise in deduction

**Problem**      **Part I**

A science student heated some ammonium chloride crystals in a test-tube and obtained an orange sublimate.

On further examination of the test-tube, a grey, dusty film was observed and the student recalled using the test-tube earlier as a makeshift spatula to remove some iron filings from a bottle.

What could have happened to the ammonium chloride crystals?

**Part II**

It was established by the student that the orange material was subliming.

The student also suspected that the reaction was concerned with iron(III) oxide. This theory was tested as follows:

A small amount of iron(III) oxide was heated with excess ammonium chloride. More orange sublimate of a stronger colour was produced than with the iron filings. As the heating continued, the residue at the bottom of the test-tube evaporated, leaving behind a thick deposit of orange sublimate higher up the tube.

The reaction was:

$Fe_2O_3 + 6NH_4Cl \rightarrow 2FeCl_3 + 6NH_3 + 3H_2O$

What did this reaction do to the iron?

How could this be useful in industry?

**Prior knowledge**
Reactions and properties of iron and ammonium chloride.

**This problem is suitable for:**
■  First year students
■  A tutorial – group work (three or four students)
■  Approximately 1 hour

**Knowledge/skills gained**
Reactions with iron and ammonium chloride.
Improved group working skills.

T46

# Orange sublimate – an exercise in deduction

Part I should be given to the students first. On completion of this Part II can be issued.

The aim of this problem is for students to discover the reactions and then reveal why ammonium chloride is used as a flux in soldering.

**Solution**

### Part I

Students should realise that ammonium chloride has reacted with iron.

Students should be encouraged to think what colour the ammonium chloride crystals would originally have been (white), and the properties of ammonium chloride. Focus their thoughts on what happens when ammonium chloride is heated (the clue is in the question).

What happens when ammonium chloride is heated? It sublimes. This is due to its dissociation into ammonia gas and hydrogen chloride gas.

$$NH_4Cl_{(s)} \rightarrow NH_{3\,(g)} + HCl_{(g)}$$

What happens when iron reacts with the hydrogen chloride gas?

$$Fe_{(s)} + 2HCl_{(g)} \rightarrow FeCl_{2\,(s)} + H_{2\,(g)}$$

As the equation above shows, iron(II) chloride is produced. The students should identify that iron(II) chloride is coloured and since it is ionic it should not sublime.

This could be deduced from the results of asking what properties iron(II) chloride has. It therefore seems logical that students then go on to think about iron(III) chloride.

Iron(III) chloride seems to be a lot more likely, as it is orange when hydrated and with its covalent character it could possibly sublime when heated.

At this point students should be asked:
How iron(III) chloride is normally made? (By reacting iron metal with chlorine gas, not hydrogen chloride gas)

Now that a problem has occurred it is necessary to try a new train of thought. What can the students say about the iron? They should, with some brainstorming and a small amount of prompting, deduce that iron(III) oxide on the surface of the iron could be the true reactant. When the students have got this far you can give them part 2 of the problem.

**Part II**

What did the reaction do to the iron and how could this be useful in industry?

The oxide layer is removed and the reaction explains why ammonium chloride is a useful flux in soldering and brazing. The word flux means flow and the fact that the reaction products melted nicely explains the term. With the oxide layer removed, the solder or brass can stick to the bare metal parts being joined.

**Acknowledgement**

Dr Cedric Mumford, University of Wales Institute, Cardiff

## S47

# Lubricating oil in milk powder

**Problem**

An industrial company which makes food products based on milk powder had some problems with their mixing machinery.

The company are concerned that some of the lubricating oil may have gotten into the milk powder. They need to know the answers to the following questions:

1. What is the lubricating oil?
2. Has any of it got into the milk powder? If so, how much?
3. Will it do the consumer any harm?

The company has supplied a bottle of the oil, a sample of pure milk powder, and a sample of the powder that is thought to be contaminated.

**Prior knowledge**
The problem relies a lot on recall and prompting by the tutor. Knowledge is needed about the constituents of foods, structures of fats and their reactions, solubilities and some spectroscopy.

**This problem is suitable for:**
■ Second year students
■ A tutorial – group work and class discussion
■ Approximately 1–2 hours

**Knowledge/skills gained**
The problem will improve the ability to think clearly and logically, to devise analytical procedures and to recognise the implications of choices.

**Butter**

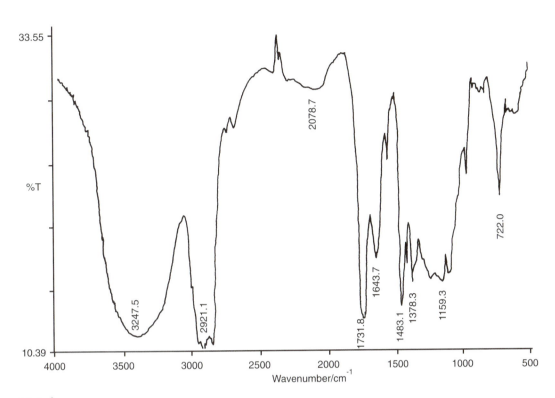

**Nujol**

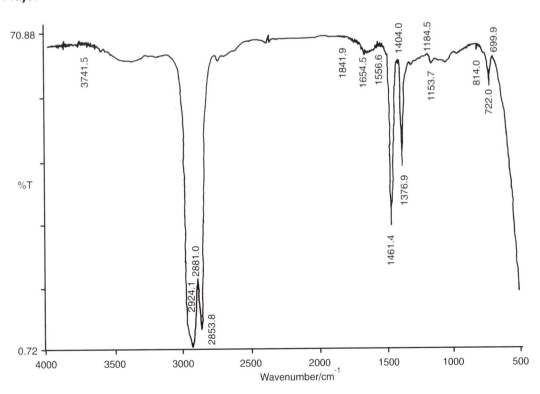

# Lubricating oil in milk powder

**Solution**

The following questions, with answers, may be helpful but they should be given out only when the students appear to have difficulties with the problem.

Suggested questions for part 1 of the problem:

Some students may suggest finding the COSHH report for the lubricating oil.

■   What properties would it be useful for the lubricating oil to have?
Slippery, viscous, high boiling point, chemically inert, cheap.

■   What sort of compound would meet these requirements?
Aliphatic hydrocarbons of longish chain length.

■   How would we demonstrate that structure?
If spectroscopy is suggested, especially IR and/or NMR, show students and see if the link with Nujol is made. If not they can be guided to it.

Suggested questions for part 2 of the problem:

■   How can we show its presence in the milk powder?
A digression will ensue to establish what milk powder is: proteins, carbohydrates and fats should be identified as the main ingredients and their chemical structures and properties can be noted. They should come to recognise that separation is necessary.

■   How can we separate the oil?
Guide them to solvent separation. What dissolves in what? What is a good solvent for the oil? If progress is fast, the need for redistilled solvents can be raised. Which component of the milk powder will also dissolve in a petrol solvent?

■   How can the oil be separated from the milk fat mixed with it?
What is milk fat? How can it be converted into water-soluble components?

■   How much are we looking for?
What quantities are commonly used in a baby's feed or in a dietary drink? How much of the oil can be conveniently quantified/determined? How?

If the students come up with alkaline hydrolysis to see if there is any unhydrolysed material remaining which should be the lubricating oil, give the following result:

Give a mass for unhydrolysed material and show the IR spectrum of butter. This completes the answer to the second part.

Suggested answers to part 3 of the problem:

The third part may be answered by informing the students that Nujol is a US brand name for liquid paraffin, used as a laxative.

If time allows, invite students to devise a method of showing that the analytical procedure works. By taking some of the sample which showed no unhydrolysed residue, spiking it with a small amount of oil and showing that the oil can be recovered. This will give confidence in the analysis and show that the method used can be repeated and still give consistent results.

■ What can be said about the purity of the milk powder sample?
Is it pure or is it just that if there was any oil it was less than a certain amount? How much?

**Acknowledgement**
Dr Stephen Breuer, Lancaster University

# Quality control of imported ethyl cyanoethanoate

**Problem**

Your company imports ethyl cyanoethanoate, $CH_3CH_2OOCCH_2CN$. Devise an analytical scheme to determine whether the product you receive is of appropriate quality.

Each consignment consists of ten 500 litre drums.

Your specification is as follows (all composition figures given in % by mass):

| | |
|---|---|
| **Identity:** | ethyl cyanoethanoate |
| **Purity:** | at least 99% |
| **Water content:** | no more than 0.5% |
| **Ethyl chloro- and dichloro- ethanoates:** | no more than 100 ppm of each |

In your answer you should consider the following, if appropriate:
■ sample collection and preparation;
■ how often and how many samples to take;
■ suitable analytical methods;
■ evaluation of your results;
■ how to ensure quality; and
■ the costs of your programme.

Ethyl cyanoethanoate is a colourless liquid of boiling point 206 °C. It is immiscible with water and hexane but miscible with methanol, ethoxyethane and propanone. It has a UV absorption maximum at 289 nm with molar absorptivity of about 2.50 $dm^3\,mol^{-1}cm^{-1}$.

**Prior knowledge**
Sampling techniques and analytical methods.

**This problem is suitable for:**
■ Second year students
■ A tutorial – group work
■ Approximately 2 hours

**Knowledge/skills gained**
Analytical strategies.

# Quality control of imported ethyl cyanoethanoate

A suggested outline answer has been supplied.

**Solution**

■ **Sample collection and preparation; how often and how many samples to take**

As the sample is a liquid, the contents of each drum should be homogeneous unless there are immiscible layers. It is necessary therefore to sample both from the top and the bottom of each drum.

If the two samples are miscible then further analysis can be done on the combined sample. (Each drum should be weighed and the liquid level determined with a dipstick – any deviation from the theoretical mass may indicate adulterants like stones or metal).

■ **Suitable analytical methods**

Infrared spectroscopy is a useful technique to identify the contents of the samples. Run samples using NaCl plates. Firstly obtain a spectrum of pure ethyl cyanoethanoate and compare spectra of samples.

With high purities the best approach is to determine the impurities, as a far lower precision can be acceptable. For a liquid boiling at moderate temperature, gas chromatography (GC) with flame ionization detection (FID) would find all the volatile organic impurities. The FID response is only approximately constant from one substance to another, but even errors of +/-2% in this assumption will contribute less than +/-0.2% to the purity value. A check on involatile impurities can be made by evaporating a sample in a weighed container and determining any residue. This is best carried out with small (1g or so) samples in a vacuum oven.

The GC–FID determination will not find water, which will have to be determined separately. Karl-Fischer titration is the best approach here.

The determination of the chloroethanoates at a 100 ppm level would be difficult using the FID but might be possible if the column was overloaded with sample. Identification would then be possible by running a second sample spiked with an extra 100 ppm of each chloroethanoate and comparing retention times. Alternatively GC–MS monitoring of characteristic ions from these species or GC with electron capture detection would add the sensitivity and specificity. Calibration would then be against known standards and it would be best to find a suitable internal standard.

■ **Quality assurance**

Using certified material for identity, spiked samples for identity of impurities, checking GC separation *eg* using higher resolution column (capillary) when doing first few samples and on occasional routine samples. In addition, normal checks should be made on instrument performance, equipment should be recalibrated regularly and staff trained.

RS•C

■   **Costs**

**Capital costs**

| | |
|---|---|
| Capital GC-FID | up to £10 000 |
| IR | £10 000 + |
| Karl-Fischer | £2000–5000 |
| GC-ECD | £10 000 |
| GC-MS | £40 000 + |

**Running costs**

Typically about £15 per determination (GC, IR, KF) – perhaps more for
GC–MS plus sampling and reporting. Overall £500–£1000 per batch.

**Acknowledgement**

Dr Colin Peacock, Lancaster University

## S49

# Hydrocarbon levels in the atmosphere near petrol stations

**Problem**

One of the major sources of hydrocarbons in the urban atmosphere is from petrol filling stations. How can you regularly monitor the hydrocarbon levels from a large town which has about one hundred stations, each to be sampled at least once per week? The limit of detection should be commensurate with the recommended air quality limit of 0.2 ppm hydrocarbons in air.

In your answer you should consider the following:
■ sample collection and preparation;
■ how often and how many samples to take;
■ suitable analytical methods;
■ evaluation of your results;
■ how to ensure quality; and
■ the costs of your programme.

**Prior knowledge**
Sampling techniques and analytical procedures.

**This problem is suitable for:**
■ First year students
■ A tutorial – group work
■ Approximately 1 hour

**Knowledge/skills gained**
Analytical strategies.

# Hydrocarbon levels in the atmosphere near petrol stations

An approach to the problem is outlined below.

**Solution**

There is a choice between on-site measurement and collecting the sample and taking it away for determination.

On-site measurement would be preferable. On-site techniques would have to be robust. Possibilities are infrared spectroscopy with a long pathlength cell; gas chromatography with flame ionisation detection (FID) or photoionisation detector (PID); FID without pre-separation. FID has drawbacks since gas cylinders would be required, but they can be small enough for up to 1 day's running on portable equipment. For hydrocarbons straight FID is simplest.

**Sampling** – small compressor to pass air continuously into FID. A sniffer head could be used to look at different areas in the station.

**Calibration** – with Certified Reference Material (CRM) gas bottles.

**Validation** – with independent Standard Reference Material (SRM) gas samples.

**Reporting** – PC or similar to log all locations, times, and data to Good Laboratory Practice (GLP) type standards. Laboratory Information Management Systems (LIMS) to compare different operators and equipment.

**Cost** – the instrument costs a few thousand pounds. One person operation.

**Frequency of sample collection** – states at least once a week in the problem, any more would be beneficial but is not necessary to meet requirements.

**Number of samples** – considering that this analysis is being performed on a large town, the number of samples taken should be consistent with the amount of time allowed.

**Acknowledgement**
Dr Colin Peacock, Lancaster University

## S50

# Investigation of a contaminated factory site

**Problem**

Summarise an appropriate analytical procedure for tackling the following problem:

A derelict factory site of about 2 hectares is to be reclaimed for a housing development. Local residents claim that large quantities of oil-based materials used to be stored on this site. Suggest a method for surveying the site to find areas of contamination by oil and to check for the presence of polychlorinated biphenyls (PCBs) and for polyaromatic hydrocarbons (PAHs)?

Your answer should consider the following:
■ sample collection and preparation;
■ how many samples to take;
■ suitable analytical methods;
■ evaluation of your results;
■ how to ensure their quality; and
■ the costs of your programme.

**Prior knowledge**
Sampling techniques and analytical methods.

**This problem is suitable for:**
■ Second year students
■ A tutorial – individual or group work
■ Approximately 2 hours

**Knowledge/skills gained**
Sampling strategies; appreciation of the value of analytical skills.

# Investigation of a contaminated factory site

This problem could be used for individual assessment or group work. Group work would probably be most beneficial as there is a lot of scope for discussion.

Suggestions have been supplied below for some approaches that could be followed.

**Solution**

### ■ Sample collection and preparation

Old plans or pictures of the site would show areas where contamination may be most likely.

As 2 hectares is approximately 100 m x 200 m then sampling on a grid of 10 m$^2$ would give 200 samples and a grid of 5 m$^2$ would give 800. Exploring the site may show areas of discolouration that could be sampled at closer intervals than apparently clean areas.

For a preliminary analysis, soil samples could be taken from just under the soil using a soil auger. Each sample should be labelled with its grid reference, the time and date and who took the sample.

### ■ Suitable analytical methods and evaluation of results

Initial samples should be quickly screened; dried at <40 °C; roughly sieved to get the <1 mm fraction; extracted with $CH_2Cl_2$ and the weight of extracted material found.

Areas showing high soluble content can then be examined in more detail. Any anomalous results, *ie* those out of line with neighbouring ones, can also be repeated.

Suspect areas should be examined in more detail, especially at various depths. Boreholes, pits or trenches may be dug to get samples down to 1m or so. Again these samples can be screened for soluble content.

A few composite samples should be prepared and screened for the presence of PAHs and PCBs. If any of these are positive then the disaggregated samples will have to be analysed separately.

General screening of the soluble organics can be done by GC–FID. Diesel oil (*n*-paraffins), Luboil (HC hump), aromatics (*eg* benzene) should all be readily identifiable. GC–MS, if available, would be very useful here.

PAHs need cleansing from bulk hydrocarbons by solid phase extraction and then characterising by HPLC/fluorescence.
PCBs need cleansing similarly and then characterising by GC–ECD.

## ■ Quality

Known blank and standard (SRM) samples should be included in the
extractions and determinations to provide a check on false positives
(contamination) and false negatives (lack of analytical sensitivity).
Quantification of PCBs and PAHs would be done using attested reference
standards.

## ■ Costs

Most of the costs will be labour costs. The cost of a professional analyst
working in a well-equipped laboratory is approximately £60 per hour. Initial
site research could take 40 man-hours. To collect, dry and extract samples
would be at least 30 minutes per sample, so any initial study that cut down
the number of samples is worthwhile. Assuming in the end about 400
samples for screening means about £30 each; complex work up for PAHs
*etc*, £50 each. If the site is at all contaminated then at least 20 such samples
would need to be done. Overall cost is then between £7000 if the site was
hardly contaminated, up to £15 000 for a few contaminated areas and
£25 000 if much of the site was suspect.

### Acknowledgement
Dr Colin Peacock, Lancaster University

S51

# Monitoring the effluent from a disused domestic waste tip

**Problem**

Summarise an appropriate analytical procedure for the following problem:

A scheme for monitoring the effluent from a disused domestic waste tip must be set up. Ground water draining from the tip must be monitored for heavy metals and acidity. In addition, it is hoped that the gaseous emissions could be burnt to provide some on-site electrical generation. The emissions should therefore be analysed for methane, carbon dioxide, nitrogen and hydrogen sulfide.

Your answer should consider the following:
■ sample collection and preparation;
■ how often and how many samples to take;
■ suitable analytical methods;
■ evaluation of your results;
■ how to ensure their quality; and
■ the costs of your programme.

**Prior knowledge**
Sampling techniques and analytical methods.

**This problem is suitable for:**
■ Second year students
■ A tutorial – individual or group work
■ Approximately 2 hours

**Knowledge/skills gained**
Sampling strategies and appreciation of the value of analytical skills.

## T51

# Monitoring the effluent from a disused domestic waste tip

This problem could be used for individual assessment or group work. Group work would probably be most beneficial as there is a lot of scope for discussion. Suggestions have been supplied below for some approaches that could be followed.

**Solution**

■  **Sample collection and preparation**

The site should be surveyed to find groundwater run-offs/drains. Probably no more than six sources of water are needed for simple monitoring. If considerable contamination was found then all run-off water would have to be monitored to find the total pollutant output. Gas output would need to be monitored by means of boreholes drilled into the tip. Perforated pipes could then be inserted to take gas samples at different points averaged over depth. If there were no obvious groundwater sources then the boreholes would have to extend to the water table so that water samples could be taken from them.

Water samples would be collected in rigid plastic bottles. As total pollutants are of interest (particulates as well as dissolved metals) the samples do not need to be filtered on-site but can be immediately stabilised by acidifying with 1% nitric acid. This stops loss of traces by precipitation or adsorption. Samples must be clearly labelled with date and time of collection, collection site and who took the sample.

Gaseous emissions could be collected in a syringe or a gas sampling bag. Samples of 20 cm$^3$ would be ample and should be taken slowly to minimise the ingress of air.

■  **How often and how many samples to take**

Sampling would need to take place often; perhaps every week to get a feel for the variations expected and then less frequently, perhaps to as little as annually. Gas sampling would need to be done more often if gas was being drawn-off for burning, in case air was being sucked in because the draw-off was too fast.

■  **Suitable analytical methods**

As gas sampling needs to be done often, use cheaper techniques like AAS unless eg ICP/OES or ICP/MS was already available. Normal heavy metals of interest are Cu, Pb, Zn, Cd, Cr, Ni and Hg. The first six can all be identified by flame AAS at levels indicative of pollution (<0.1 ppm).

Larger samples of $H_2S$ may be required for it may be in relatively low concentration. Traces can be determined *in situ* using a 'Drager tube' (or similar) or a dissolution train collecting it in silver nitrate, cadmium chloride, or dichromate.

Water pH is best measured *in situ* using a portable meter.

Mercury is toxic at levels well below the detection limit of flame AAS so samples would have to be determined by furnace AAS or, better, by cold vapour generation AAS.

The gases can be determined by GC with a hot wire detector using a highly absorptive stationary phase like charcoal or a molecular sieve.

■ **How to ensure the quality of results**
Initially several samples should be quantified by standard addition to check interferences.

Standards should then be made in appropriate matched matrices. Methods should be checked by including standard water samples and blanks in with the samples.

■ **Cost**
Surveying the site, finding suitable water sampling points and getting the gas sampling boreholes will be quite time-consuming. It costs at least £60 per hour to support a professional analyst at work.

Sample collection is similarly time-consuming if travelling is taken into account. However, it can be left to skilled technicians as can *in situ* pH measurement. Flame AAS is a fast and cheap technique costing about £5 per determination and capable of being automated or run by relatively inexperienced staff. Cold vapour determinations are more complex but are often highly automated. Similarly GC of gases will be relatively easy, costing about £15 per sample. Expect overall about five to ten thousands pounds to set up the study and then £150 for each set of gas samples (monthly) and £300 for each set of water samples (quarterly).

**Acknowledgement**
Dr Colin Peacock, Lancaster University

# Exposure of people to phthalate esters

**Problem**

There is some evidence that phthalate esters (esters of benzene-1,2-dicarboxylic acid) can mimic oestrogenic hormones and may be a cause of the drop in sperm count amongst young adult males. Devise a scheme for a final year undergraduate project to identify the extent to which people are exposed to these substances.

Those commonly used are the diethyl, dibutyl, dicyclohexyl, dioctyl, dinonyl and dibenzyl esters, but others may be possible. The longer chain alkyl groups may be straight or branched chains and the attachment of the oxygen may be at a variety of places.

Esters are normally liquids, are readily soluble in polar organic solvents and have boiling points of up to 300 °C. They have a UV maximum around 270 nm with molar absorptivity about 10 $dm^3\,mol^{-1}\,cm^{-1}$.

In your answer you should consider the following, if appropriate:
- sample collection and preparation;
- how often and how many samples to take;
- suitable analytical methods;
- evaluation of your results;
- how to ensure quality; and
- the costs of your programme.

**Prior knowledge**
Sampling techniques and analytical methods.

**This problem is suitable for:**
- Second year students
- A tutorial – group work and class discussion
- Approximately 2 hours

**Knowledge/skills gained**
Analytical strategies.

# Exposure of people to phthalate esters

A possible solution has been outlined below.

**Solution**

As all phthalate esters are reasonably volatile, GC is the best analytical method. They will be extracted from fairly complex matrices so a specific detection method would make determination much easier. All phthalate esters, except methyl, give a MS ion at $m/z$ 149, which is quite distinctive so GC–MS would be the method of choice.

Suggestions of extraction, hydrolysis and determination of the phthalic acid by ion chromatography with UV detection should be given credit. This has the advantage of giving a sum of phthalates rather than a lot of individual esters.

### ■ Sampling

There is an annual government survey on household consumption (regularly reported in newspapers). This and other literature should be studied to determine target diets. Taking small samples from the regular meals of student colleagues could give a useful range of samples, especially if it could include, for example, vegetarians, fast food addicts.

### ■ Suitable Analytical Methods

For the experimental method some discussion along the following lines would be expected:

**Contamination:** phthalates are so ubiquitous that sampling and storage containers, solvents and apparatus must be carefully checked.

**Extraction:** homogenised samples of representative diets would need extracting, probably by Soxhlet, into the least polar solvent to leave as much extraneous material behind as possible (*eg* hexane and methanol). If hexane is a very poor solvent than an initial extraction with hexane followed by a further one with a more polar solvent may give some clean-up.

**Clean-up:** adsorption onto silica and elution with methanol is one possibility. Use of solid phase extraction cartridges may be possible (if the £1–£3 per cartridge can be borne).

**Recovery tests:** will be needed on a variety of matrices and for a range of phthalates – checking spiked samples extracted against samples extracted and then spiked.

### ■ Standards

If using GC–MS, standard samples will be needed to ascertain response factors and a suitable internal standard found. The ideal internal standard would be an isotopically modified (*eg* deuterated) phthalate which could be added when the sample is being homogenised and be taken through the whole extraction cycle.

■ **Costs**

GC–MS capital cost £40 000+

Running costs would be high-purity solvents (£10–40 per litre) and clean-up materials. Labour costs would be high if this was being done commercially (£60 per hour, so perhaps £300 per sample).

**Acknowledgement**

Dr Colin Peacock, Lancaster University

# Quality control of aspirin manufacture

**Problem**

How would you monitor, for manufacturing quality control purposes, the aspirin content of effervescent tablets of the following formulation:

| | |
|---|---|
| Aspirin, sodium salt | 300 mg |
| Fructose | 100 mg |
| Sodium hydrogencarbonate | 100 mg |
| Citric acid | 50 mg |
| Binder (modified starch) | 30 mg |

The formulation has to be manufactured with a limit of +/-15 mg on the aspirin content. Production is carried out in 50 kg batches, generally one a day.

In your answer you should consider the following:
- sample collection and preparation;
- how often and how many samples to take;
- suitable analytical methods;
- evaluation of your results;
- how to ensure their quality; and
- the costs of your programme.

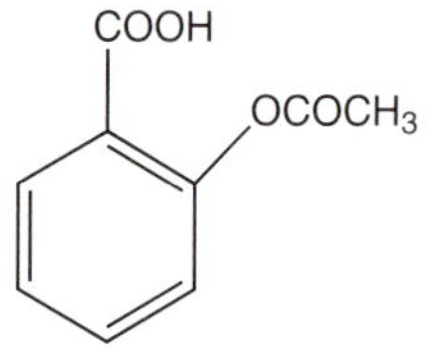

Structures for aspirin and citric acid are shown below:

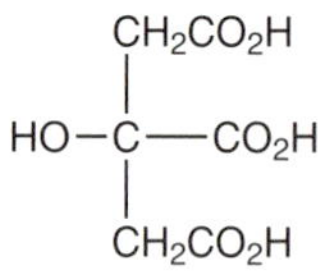

Aspirin (2-ethanoyloxybenzoic acid)

Citric acid (2-hydroxypropane-1,2,3-tricarboxylic acid)

**Prior knowledge**
Sampling techniques and analytical procedures.

**This problem is suitable for:**
- Second year students
- A tutorial – group work
- Approximately 1 hour

**Knowledge/skills gained**
Analytical strategies.

# T53

# Quality control of aspirin manufacture

A suggested outline solution to the problem is given below.

**Solution**

■ **Suitable analytical techniques**

The components of the tablets are mainly water-soluble, but aspirin can be taken into ethoxyethane if acidified. Various techniques could be used: NMR (but to achieve an accuracy of better than +/-1% in analysis would be difficult); IR in solution (but aqueous systems cannot be used). The components after acidification would have to be extracted into an organic solvent such as trichloromethane, which would probably be the best, and then run as liquid sample; UV – only aspirin has an aromatic chromophore. Dissolve in water and record absorbance (preferably at more than one wavelength in case of interference); HPLC or ion chromatography with UV detection would be certain to allow separation from excipients.

■ **Standards**

The student needs to show that the assay is precise – *ie* that calibrations are linear over a range rather greater than the expected variation in manufactured composition. At least ten independent samples will be needed if the results are to be valid.

The readings should be calibrated against CRM (Certified Reference Material) aspirin and the method should be checked against regular in-house reference samples to make sure that it is still working.

Results should be reported to GLP (Good Laboratory Practice) standards. Ideally all data should be recorded on computer – sample reference, operator, masses, analytical results, calculations, as should the final report. Data can then be used to follow the performance of laboratory equipment and operators.

**Acknowledgement**
Dr Colin Peacock, Lancaster University

# Mercury in sediments

**Problem**

A research project is to be undertaken to study the speciation of mercury in sediments.

Consider fully the problems which will be encountered in devising and developing suitable methods for the quantitative analysis of the mercury.

**Prior knowledge**

The chemistry of mercury; atomic absorption spectroscopy.

**This problem is suitable for:**
■ Second year students
■ A tutorial – group work
■ Approximately 1 hour

**Knowledge/skills gained**

Application of a spectroscopic technique for analysis. Treating a sample to preserve it for analysis.

# Mercury in sediments

A suggested approach is given below.

**Solution**

The study of the heavy metals is an important part of analytical chemistry. Mercury is a highly toxic element and its levels in the environment are issues that cause concern. Mercury occurs naturally, for example in unpolluted soils where it probably originated from the weathering of parent rock. However, it can also occur in the soil from pesticides, smelting or aerosol deposition.

When storing samples containing mercury for analysis, care has to be taken to avoid possible loss of mercury and resultant contamination. Mercury could be lost by reduction of Hg(II) to Hg or by conversion into organomercury compounds. It has been shown that Teflon containers can reduce the Hg(II) ions, therefore samples should be stored with oxidising agents.

To give a total mercury content it is normally essential that the sample is decomposed to produce Hg(II) in solution. To ensure there is no loss of the metal during this procedure a Schroniger flask is used. The sample is ashed in oxygen in the system, allowing for any increase in pressure. The ashing is performed in an acidic solution *eg* $H_2SO_4$ with an added oxidising agent *eg* $KMnO_4$.

The analysis of the mercury can be done using several techniques. However, the one that is most widely used is Atomic Absorption Spectroscopy (AAS) with cold vapour. The excess potassium manganate(VII) is destroyed by hydroxylamine, and a reducing agent *eg* $SnCl_2$ is used to reduce the mercury ions to mercury. The mercury is flushed out by bubbling an inert gas such as nitrogen through the solution and the vapour is swept into a cell in the AAS.

**Acknowledgement**
Based on an idea from Andy Gray, Solvay Interox Ltd, Widnes, Cheshire

## S55

# Contamination of pharmaceutical manufacturing equipment

**Problem**

Machines used to manufacture tablets and capsules need to be cleaned between the production of batches. This is to prevent contamination of one product with the next. Discuss the swabbing of the equipment in order to check its cleanliness. Some examples of ideas to consider include when and where the cleaning should occur, and with what.

**Prior knowledge**
Swabbing techniques.

**This problem is suitable for:**
■   Second year students
■   A tutorial – group work and class discussion
■   Approximately 1 hour

**Knowledge/skills gained**
The importance of clean equipment for manufacturing.

## T55

# Contamination of pharmaceutical manufacturing equipment

Manufacturing equipment must be clean to prevent cross contamination when switching from one product to another. A suggested approach for this problem is given below.

**Solution**

Some suggested questions and answers to this problem are listed below. These questions could be asked outright by the tutor with the students responding with their thoughts, or the problem could be made even more discussion-orientated by the tutor leaving all points for discussion open, that is, not supplying the questions.

■ **When do you swab the equipment?**
After manufacture of a product is finished there will be a clean-down. If the equipment after this initial clean-down can be seen by the naked eye to be contaminated then it will fail and will need to be cleaned again. If it looks clean go ahead and swab.

■ **Where do you swab?**
Choose a representative area of the machine to swab, not the whole equipment. Choose a place likely to have remained contaminated.

■ **What do you use to swab?**
Suggestions would be cotton wool, foam, tissue, cloth. You need to be able to recover the active compound from the swab material.

■ **What solvent do you use to swab with?**
The active compound must be soluble in the solvent. The toxicity of the solvent needs to be considered. Is any residue from the solvent left on the equipment? Any possible corrosion of the equipment by the solvent must also be considered.

■ **What volume of solvent do you swab with?**
Keep the volume small – eg 10 cm$^3$ – so that low levels of active compound are likely to be detected.

■ **Choose an appropriate method of analysis for swab solutions**
HPLC is the technique of choice due to its specificity, sensitivity, speed and availability of methods to determine the active compound. It is necessary to be able to validate the method. How easy is it to recover active compounds from swab material? How stable is the swab solution? What effect does filtration have on the active solution?

■ **Permissible level of residual active compound on equipment before reuse**

The permissible level is related to the dosage of active compound.

**Acknowledgements**

Mrs Alison Bretnall and Dr Thomas Cowen, Bristol-Myers Squibb Pharmaceutical, Moreton, Wirral